SAFETY FOR WELDERS

Safety for Welders

Larry F. Jeffus

Library of Congress Control Number: 78-73579

ISBN-13: 978-0-8273-1684-3

ISBN-10: 0-8273-1684-4

Delmar Cengage Learning
5 Maxwell Drive
Clifton Park, NY 12065-2919
USA

Cengage Learning products are represented in Canada by Nelson Education, Ltd.

For your lifelong learning solutions, visit **delmar.cengage.com**

Visit our corporate website at **www.cengage.com**

Notice to the Reader
Publisher does not warrant or guarantee any of the products described herein or perform any independent analysis in connection with any of the product information contained herein. Publisher does not assume, and expressly disclaims, any obligation to obtain and include information other than that provided to it by the manufacturer. The reader is expressly warned to consider and adopt all safety precautions that might be indicated by the activities described herein and to avoid all potential hazards. By following the instructions contained herein, the reader willingly assumes all risks in connection with such instructions. The publisher makes no representations or warranties of any kind, including but not limited to, the warranties of fitness for particular purpose or merchantability, nor are any such representations implied with respect to the material set forth herein, and the publisher takes no responsibility with respect to such material. The publisher shall not be liable for any special, consequential, or exemplary damages resulting, in whole or part, from the readers' use of, or reliance upon, this material.

Printed in the United States of America
22 23 24 25 11 10 09 08

This book is dedicated to my wife, Mary
and my family for all their help and support.

Preface

This book presents a collection of safety habits and precautions that can be helpful for those who work in a welding shop. Many general and specific safety rules are suggested.

SAFETY FOR WELDERS is designed for beginning students of welding. Little or no prerequisite knowledge of welding is needed to enable thorough understanding of the content or to benefit from the safety rules presented.

The text consists of ten units that discuss welding safety in brief, easy-to-read guidelines which lend themselves to independent study on the part of the student. Each unit is followed by review questions that provide an opportunity for students to test their own newly-obtained knowledge.

Accident prevention for students is the main intent of SAFETY FOR WELDERS. However, the safety messages included must be treated only as general guides. Because of the variations in human judgment, personal ability, mechanical aptitude, and existing conditions, these messages should not be accepted as foolproof guarantees against personal injury. There is no substitute for caution and common sense. If students have any doubt about safety measures, they should consult their instructor before operating any equipment or beginning a procedure.

There are in existence other sources of safety codes for the welding industry. When the instructor or welder has any doubt about safety conditions, a code should be consulted. Several of these code sources are listed, many others are available:

American Welding Society
2501 N.W. 7th Street
Miami, Florida 33125

National Electrical Manufacturers Association
155 East 44th Street
New York, New York 10017

National Safety Council
425 North Michigan Avenue
Chicago, Illinois 60611

Occupational Safety and Health Administration
U.S. Department of Labor
Washington, D.C. 20210

Underwriters' Laboratories
207 East Ohio Street
Chicago, Illinois 60611

Other welding textbooks from Delmar include:

Basic Arc Welding
Basic Oxyacetylene Welding
Basic Tig & Mig Welding
Blueprint Reading For Welders
Industrial Welding Procedures
Pipe Welding Techniques
Practical Problems In Mathematics For Welders
Welding Procedures: Electric Arc
Welding Procedures: Oxyacetylene
Welding Procedures: Mig & Tig
Welding Processes

Acknowledgments

The following people and companies have contributed to this text.

Illustration

Crouse-Hinds: 8-1
Frommelt Industries, Inc. - Safety Products Div.: 2-6
David Rhodes: 2-4, 5-1, 7-10, 8-3, 8-4, 8-5, 9-1, 10-2, 10-3
The remaining photographs were taken by the author.

Editorial

Mark W. Huth - Sponsoring Editor
Laurie Kurzon - Project Editor

Classroom Testing

The material in this text book was classroom tested at Eastfield College.

Contents

Unit 1 Handling and Storage of Cylinders

OBJECTIVE

After completing this unit, the student will be able to safely handle and store oxy-fuel cylinders.

Handling oxy-fuel cylinders, especially acetylene, can be very dangerous. For this reason, it is important to know how to safely handle and store these cylinders.

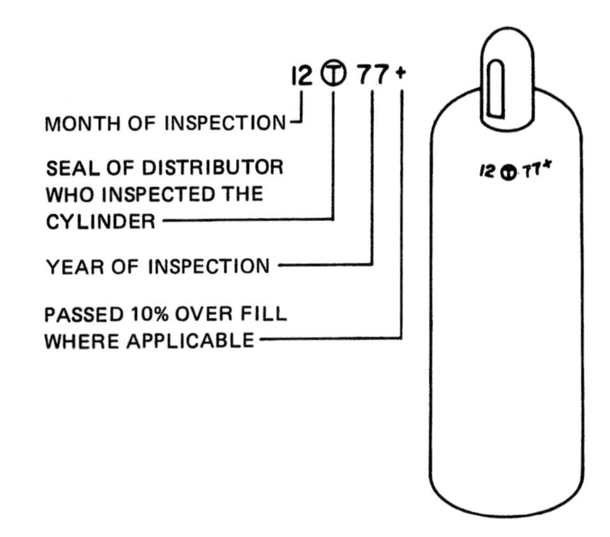

Fig. 1-1

1. CYLINDERS IN GENERAL

1.1 All cylinders must have been inspected within the last five to ten years, depending on the age of the cylinder. The U.S. Department of Transportation publishes regulations for this in 49 CFR Parts 171-179, figure 1-1.

1.2 All cylinders must be clearly marked to identify the gas content by either a chemical or trade name, figure 1-2.

1.3 Cylinder connections must comply with the American National Standard for compressed gas cylinder outlet and inlet valves. (ANSI 1357.1-1965)

1.4 Cylinders having a water-weight capacity larger than 30 pounds (13.5 kg.), must have either a means to connect a valve protection cap, or the valve must be recessed or have a collar.

1.5 Store all cylinders away from heat, such as radiators.

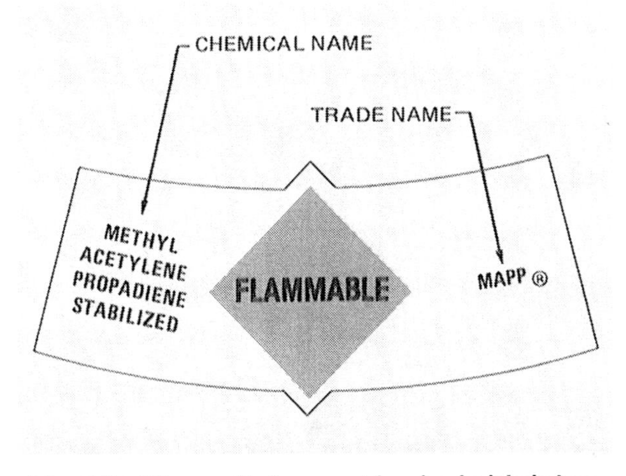

Fig. 1-2 All gas cylinders must be clearly labeled so the gas contents can easily be identified.

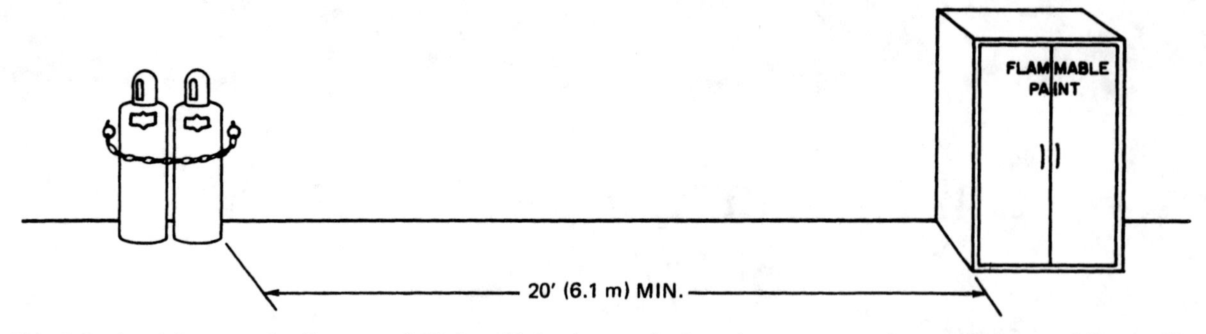

Fig. 1-3 A minimum safe distance of 20 feet (6.1 m) must be kept between stored gas cylinders and flammable materials.

1.6 Keep all stored cylinders at least 20 feet (6.1 m) from highly flammable materials such as paint, oil or cleaning solvents, figure 1-3.

1.7 Cylinders should be stored so that they will not be knocked over or tampered with by unauthorized persons.

1.8 Valve protection caps must be in place on all cylinders in storage.

1.9 All cylinders not connected to a regulator or manifold are considered to be in storage and must comply with storage rules.

1/2-HR FIRE RATING

FUEL GAS

OXYGEN

5′ MIN. (1.5 m)

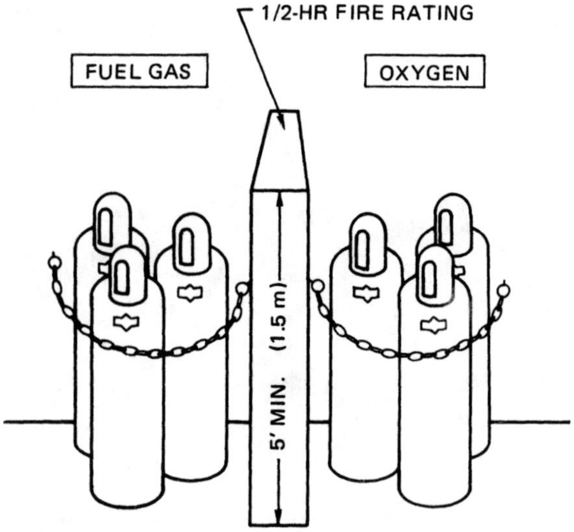

Fig. 1-4 Cylinders must be separated by a 5 foot (1.5 m) high wall if they cannot be separated by at least 20 feet (6.1 m).

1.10 Oxygen cylinders must be stored separately from fuel gas cylinders by at least 20 feet (6.1 m) or have a 5 foot (1.5 m) high, noncombustible wall with a 1/2-hour fire resistant rating, figure 1-4.

1.11 Rooms used for fuel gas storage or a manifold must be well ventilated and have a sign posted that reads: "Danger – No Smoking, Matches or Open Lights", or similar wording, figure 1-5.

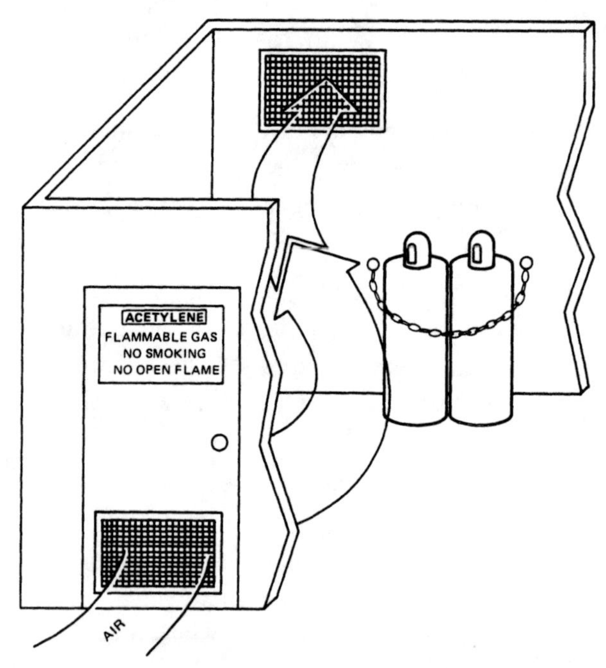

Fig. 1-5 Acetylene stored in a separate room must have good ventilation and should have a warning sign on the door.

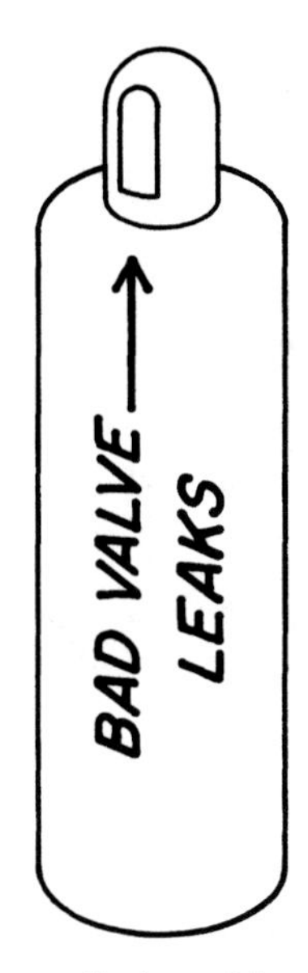

Fig. 1-6 Write on cylinder with soapstone or chalk to help suppliers identify defects.

1.12 Cylinders should be lifted by some means other than lifting by the valve protection caps.

1.13 Valve protection caps must be in place before a cylinder is moved, unless it is on a cylinder cart.

1.14 Use warm water, (not boiling), to loosen cylinders frozen to the ground.

1.15 Cylinders should be handled gently and never dropped.

1.16 Prevent cylinders from coming in contact with an electrical current, such as a welding arc.

1.17 Only gas suppliers should fill cylinders.

1.18 If a cylinder valve needs repair or cannot be opened by hand, notify the supplier and follow his instructions, figure 1-6.

1.19 Close cylinder valves on all empty cylinders before moving them.

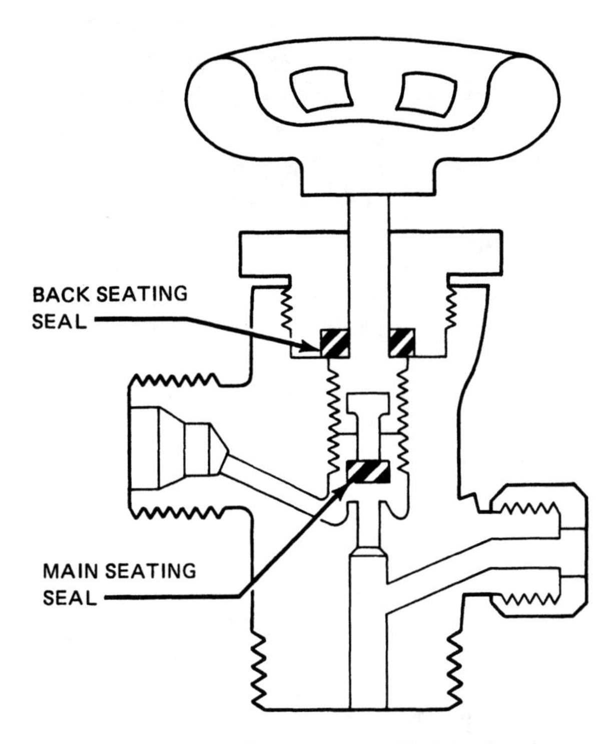

Fig. 1-7 Cutaway of an oxygen cylinder valve shows the two separate seals.

1.20 Never use cylinders as rollers.

1.21 Cylinders should always be far enough away from actual welding and cutting so that no hot slag or sparks can reach them.

2. OXYGEN CYLINDERS

2.1 Keep oxygen cylinders and oxygen equipment free of oil.

2.2 Always use oxygen with a regulator.

2.3 Never use oxygen as air to inflate tires, blow out dust, etc.

2.4 If a leaky oxygen cylinder cannot be stopped by turning off the cylinder valve, take it outside and notify the supplier.

2.5 Open oxygen cylinder valves all the way when using them, figure 1-7.

2.6 When cracking (quickly opening and closing) an oxygen valve, or any high-pressure gas cylinder valve, be sure to stand to one side. Hold the cylinder so it cannot be pushed over by the force of escaping gas, figure 1-8.

2.7 Never allow a jet of oxygen to strike greasy or oily surfaces.

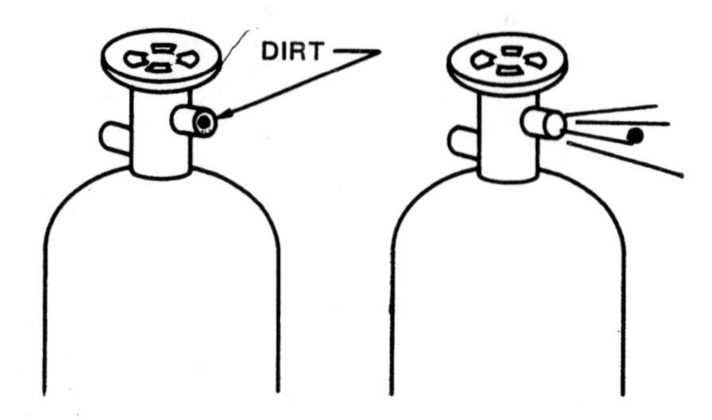

Fig. 1-8 During transportation or storage, dirt may collect in the valve. Wiping this dirt may leave oil from a hand or rag and might not remove all the dirt. Cracking the valve is the best way to remove any dirt.

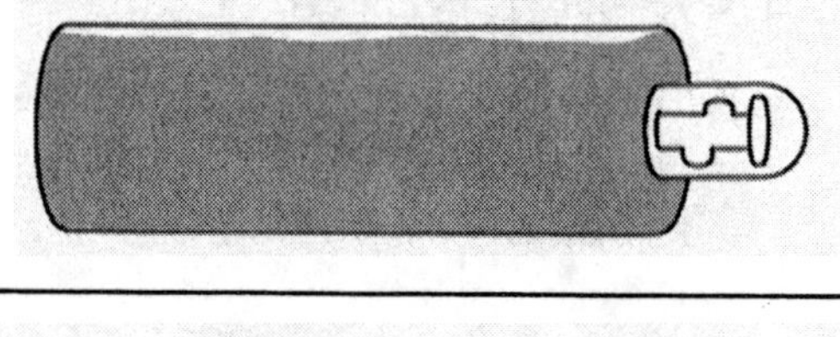

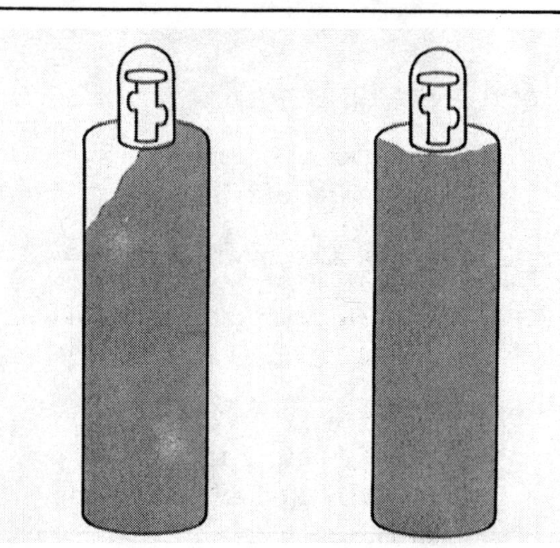

Fig. 1-9 The acetone in an acetylene cylinder must have time to settle before the cylinder can be used safely.

3. FUEL CYLINDERS

3.1 Store fuel cylinders with the valve end up.

3.2 Always use fuel cylinders with the valve end up.

3.3 If an acetylene cylinder has been on its side, set it with the valve end up for at least 15 minutes before using it. This allows the acetone to settle, figure 1-9.

3.4 Store more than 2000 cubic feet (56 m^3) total gas capacity or 300 pounds (136.2 kg) of liquefied petroleum in a separate room.

3.5 Never open acetylene cylinder valves more than 1 1/2 turns, and preferably not more than 3/4 of a turn, figure 1-10.

3.6 If a fuel cylinder has a leak that cannot be stopped with the cylinder valve, take it outside, well away from any source of ignition, and slowly empty it. Place a warning

Fig. 1-10 The less an acetylene cylinder valve is opened, the quicker it can be turned off in an emergency.

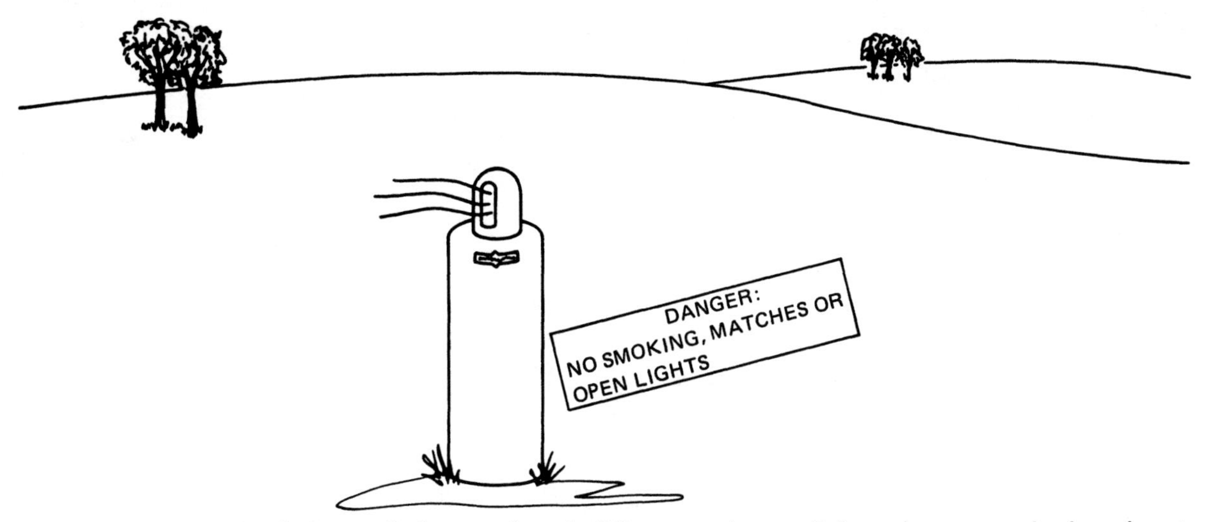

Fig. 1-11 Move a leaking fuel-gas cylinder out of any building or work area. Release the pressure slowly and post a warning sign.

near the container. Notify the supplier and follow his instructions regarding the return of the cylinder, figure 1-11.

3.7 When cracking a fuel-gas cylinder valve, stand to one side and be sure there is no open flame that can ignite the escaping gas.

REVIEW QUESTIONS

Read each question and indicate the number of the rule in the unit which most accurately answers the question.

EXAMPLE: How far should you open an oxygen cylinder valve when using it?

Answer: 2.5 In the text, 2.5 says: "Open oxygen cylinder valves all the way when using them."

1. Why should oxygen and fuel-gas cylinders not be left on a portable cart without having the regulators connected?
2. What should be done with a fuel-gas cylinder leak that cannot be stopped?
3. How often must gas cylinders be inspected?
4. How can cylinders that are frozen to the ground be loosened?
5. Where should you stand when cracking an oxygen cylinder valve?

Read each question and indicate the letter next to the statement that most accurately answers the question.

1. The acetylene cylinder valve must never be opened more than how many turns?

   a. all the way
   b. 3/4 of a turn
   c. 2 turns
   d. 1 1/2 turns

2. Oxygen cylinders must be stored at least how many feet from fuel-gas cylinders?

   a. 15 feet (4.6 m)
   b. 25 feet (7.6 m)
   c. 20 feet (6.1 m)
   d. 10 feet (3.1 m)

3. If an acetylene cylinder has been on its side, how long must it stand up straight before being used?

   a. 10 minutes
   b. 15 minutes
   c. 1 hour
   d. no waiting time needed

4. How far away from the work should cylinders be?

   a. 25 feet (7.6 m)
   b. so that hot sparks cannot reach them
   c. so that they can be reached quickly in an emergency
   d. 20 feet (6.1 m)

5. What information is required on a cylinder?

   a. type of gas contained
   b. size of cylinder
   c. pressure of the gas inside
   d. all of the above

Unit 2 The Welding Environment

OBJECTIVES

After completing this unit, the student will be able to explain the ventilation needed for a welding shop. The student will also be able to explain how to prevent ultraviolet light from injuring the welder or other people in the shop.

All welding gives off fumes. Some of these fumes can be harmful to the welder if the area is not ventilated properly.

1. VENTILATION

1.1 All fumes should be removed before they can rise past the level of the welder's face, figure 2-1.

Fig. 2-1 The large exhaust pickup next to the arc welding table will pull exhaust away from the welder.

1.2 Respirators should be provided for each person welding in a confined area where ventilation cannot remove the fumes.

1.3 Forced ventilation is always required for welding being done on mentals containing zinc, lead, beryllium, cadmium, mercury, copper, or other materials that give off dangerous fumes.

1.4 Forced ventilation for a room should be 2,000 cubic feet (56 m^3) for each welder.

1.5 Forced ventilation is not required if the room has at least 10,000 cubic feet (280 m^3) per welder, figure 2-2.

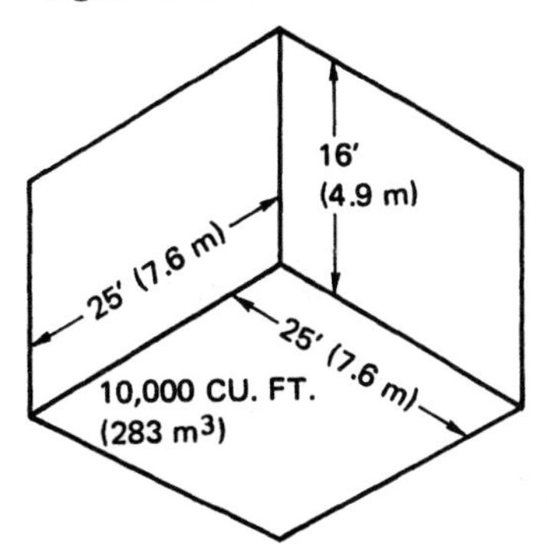

Fig. 2-2 A room 25′ x 25′ (7.6 m x 7.6 m) with a 16′ (4.9 m) ceiling has 10,000 cu. ft. (283m^3) and is large enough to have one welder without forced ventilation. The size of the room must increase proportionately to allow for more welders.

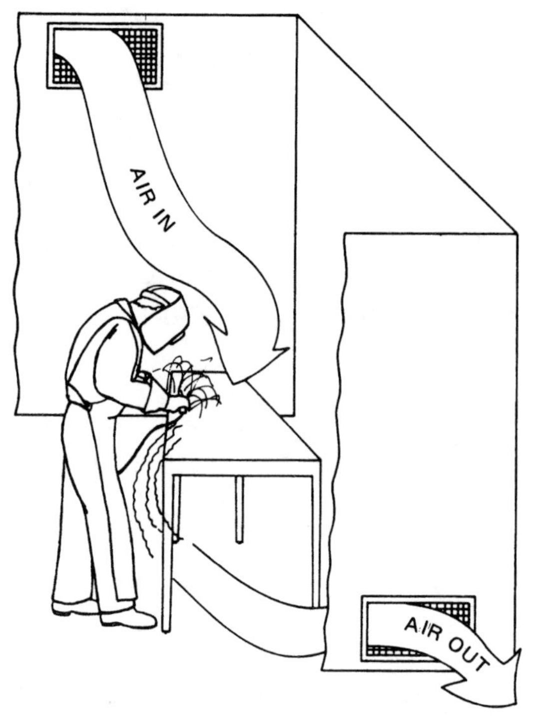

Fig. 2-3 Forced ventilation should pull the welding fumes away from the welder.

1.6 Forced ventilation is not required if the room has a ceiling 16 feet (4.88 m) or higher.

1.7 A welding booth (any fixed enclosure for welding that has two or more sides and a top) should have an air flow of at least 100 linear feet (30.5 m) per minute away from the welder, figure 2-3.

1.8 Movable exhaust pickups must have a velocity of 100 linear feet (30.5 m) per minute away from the welder in the welding zone, figure 2-4.

1.9 Extra ventilation is needed for brazing because some brazing rods contain zinc.

1.10 Welding and cutting on galvanized metal releases zinc oxide, which must be removed.

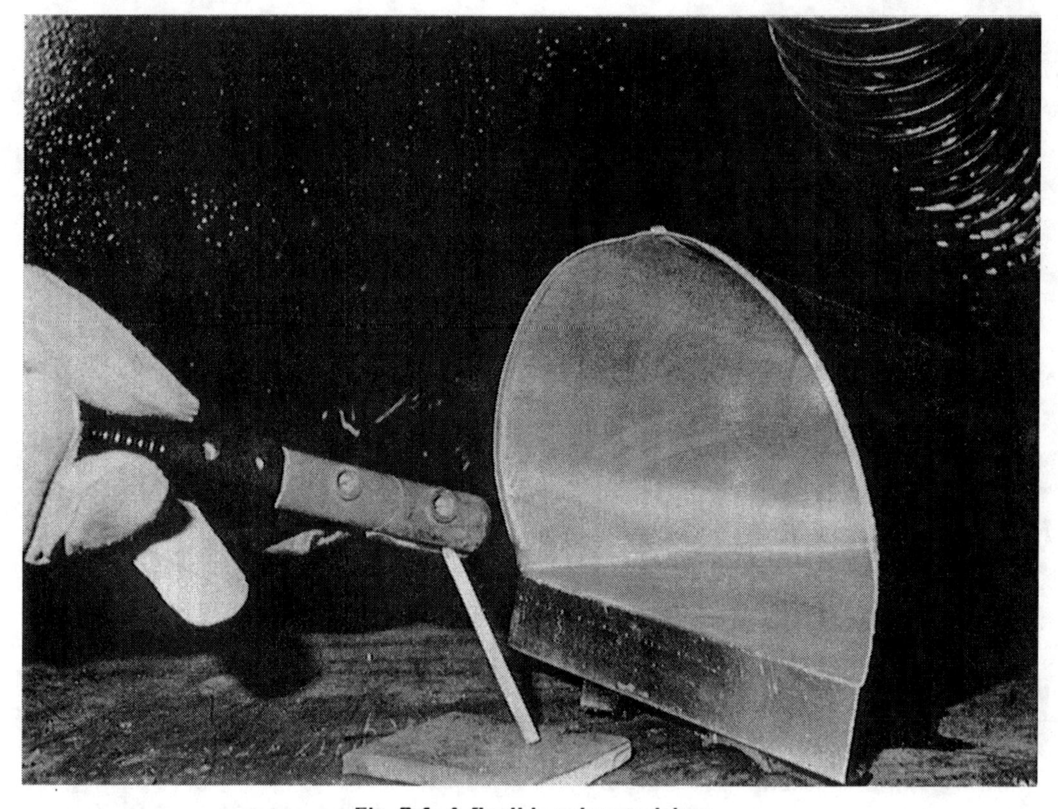

Fig. 2-4 A flexible exhaust pickup.

1.11 Welding or cutting on painted metal can release lead oxide because many paints, especially primers, contain lead.

1.12 Cadmium is an alloy found in some solders. Cadmium silver solder may be as high as 95 percent cadmium. Cadmium fumes must be avoided.

1.13 Fumes from an element such as fluoride, used in many fluxes for brazing and soldering, causes eye, sinus, and skin irritations.

2. WELDING LIGHT RAYS

Almost all welding gives off light. The light can be divided into three major groups: infrared, visible, and ultraviolet. Infrared and visible light are not as dangerous as ultraviolet, which can cause serious injuries. Some infrared and all ultraviolet light rays are invisible. For this reason, the welder must try to avoid all welding light, both direct and reflected.

2.1 Paint the inside of all welding booths flat black to absorb light and not reflect it, figure 2-5.

2.2 To prevent the reflection of light, paint ceiling and ductwork flat black.

2.3 When welding outside of welding booths, put up portable curtains to protect other people in the area, figure 2-6.

2.4 The welder's skin must be completely covered. Welding light rays can cause second-degree burns (a second-degree burn is red and blistered skin). These burns, like sunburns, can progress unnoticed for sometime.

2.5 Light burns can become easily infected.

2.6 The white and the retina, or back of the eye, can be easily burned, figure 2-7.

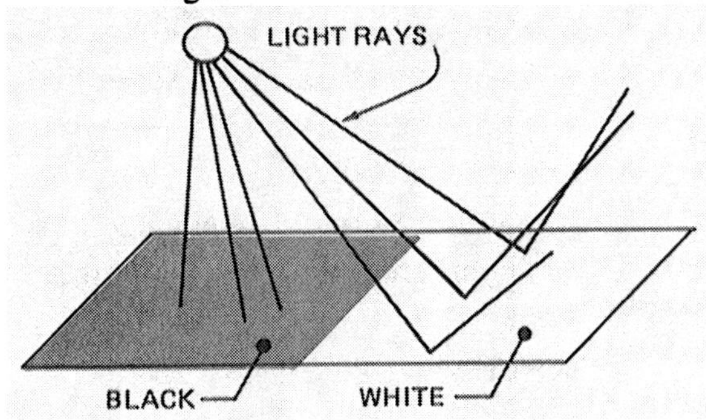

Fig. 2-5 Black absorbs light rays and white, or light colors, reflects light rays.

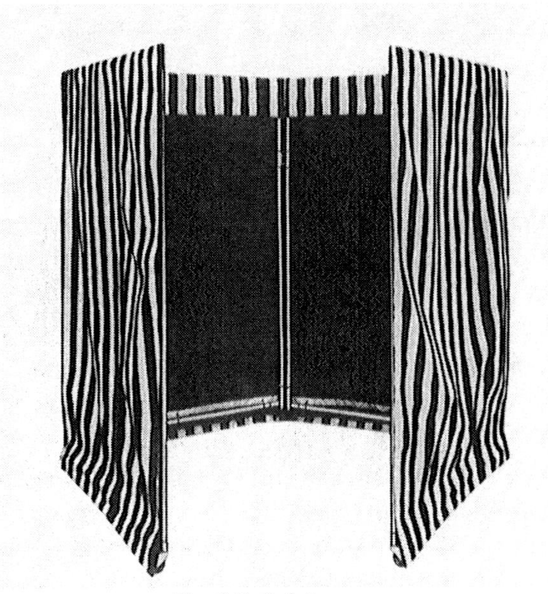

Fig. 2-6 Safety screen

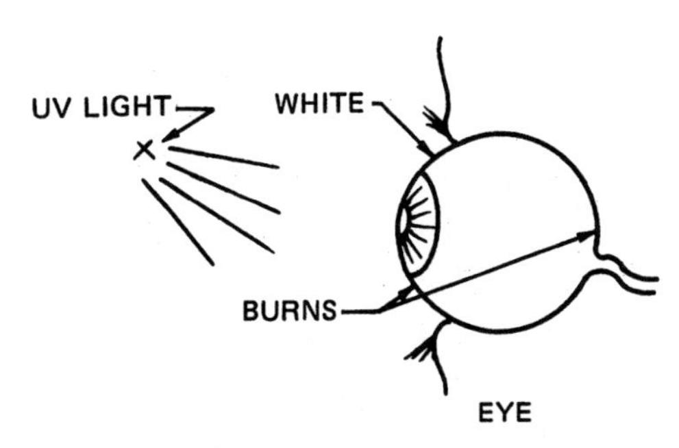

Fig. 2-7 The eye can be burned on the white or on the back by the ultraviolet light rays.

2.7 Eye burns, sometimes called welder's flash, feel like something is in the eye and may not start hurting for sometime after they happen. Because it is impossible for the welder to be sure there is nothing in the eye and there is a high chance of infection, get professional help each time there is an eye injury.

2.8 Never use another person's medicine or home remedies for eye burns.

2.9 Infrared light is felt in the form of heat. It can be as severe as ultraviolet, but its effects are felt quickly and for this reason, are more easily avoided.

2.10 Visible light is not as harmful as the others. It can cause night blindness (a temporary loss of night vision).

REVIEW QUESTIONS

Read each question and indicate the number of the rule in the unit which most accurately answers the question.

1. What should you do in case of welder's flash?
2. Why is welding or cutting on painted metal dangerous without good ventilation?
3. What should you put up when welding outside a welding booth?
4. What should be used if welding fumes cannot be removed from a confined area?
5. At what level should welding fumes be removed from the welding area?

Read each question and indicate the letter next to the statement that most accurately answers the question.

1. Forced ventilation is always needed when welding which of the following metals?

   a. titanium, aluminum, steel
   b. stainless steel, cast iron
   c. zinc, cadmium, copper
   d. aluminum, stainless steel

2. Which of the following lights is least dangerous?

   a. ultraviolet
   b. infrared
   c. visible
   d. all of the above are equally dangerous.

3. Welding light can cause what degree of skin burn?

   a. first degree
   b. second degree
   c. third degree
   d. none of the above

4. How does the welder feel infrared light?

   a. the same as ultraviolet
   b. like a sunburn
   c. it's not felt
   d. as heat

5. What element in some brazing and soldering fluxes can cause eye, sinus, and skin irritation?

   a. borax
   b. fluoride
   c. rosin
   d. none of the above

Unit 3 Fire Protection

OBJECTIVES

After completing this unit, the student will be able to explain how and where to locate fire extinguishers in a welding shop and choose the correct type of fire extinguisher for the type of fire. The student will also be able to list fire prevention techniques.

Fire is a constant danger to the welder. The welder is constantly subjecting the welding area to hot sparks from the torch or electrode. These sparks may be as hot as 2800 degrees Fahrenheit or 1538 degrees celsius and can easily ignite most combustible materials.

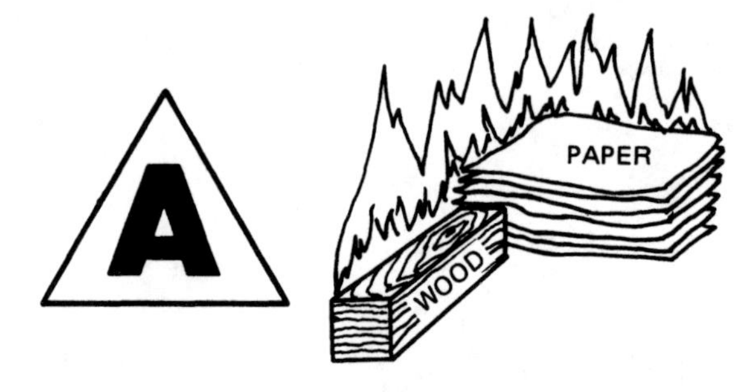

Fig. 3-1 The symbol for type A extinguishers is a green triangle with the letter A in the center. This type of extinguisher is used on flammable solids.

1. TYPES OF EXTINGUISHERS

Fire extinguishers are divided into four types. The classifications are based on the type of fire they extinguish. Some extinguishers carry two or more classifications.

1.1 Type A extinguishers are for combustible solids (things that burn), for example: paper, wood, cloth, figure 3-1.

1.2 Type B extinguishers are for combustible liquids, for example: gas, oil, paint thinner, figure 3-2.

Fig. 3-2 The symbol for type B extinguishers is a red square with the letter B in the center. This type of extinguisher is used on flammable liquids.

1.3 Type C extinguishers are for electrical fires, for example: motors, fuse boxes, welding machines, figure 3-3.

1.4 Type D extinguishers are for flammable metals, for example: magnesium, zinc, iron filings, figure 3-4.

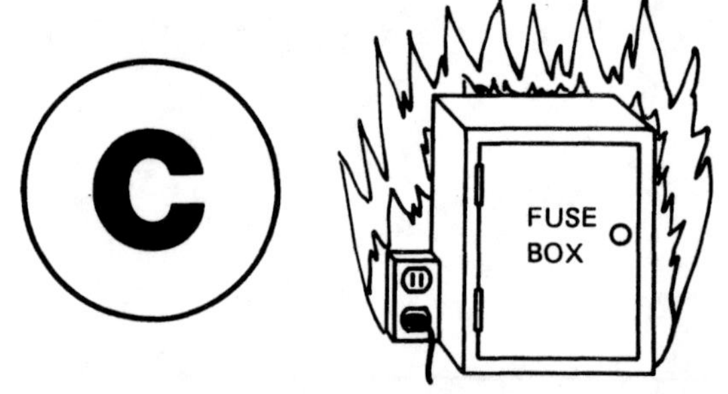

Fig. 3-3 The symbol for type C extinguishers is a blue circle with the letter C in the center. This type of extinguisher is used on electrical fires.

2. USE OF EXTINGUISHERS

2.1 Read the operating instructions on the side of the extinguisher.

D
MAGNESIUM
ZINC

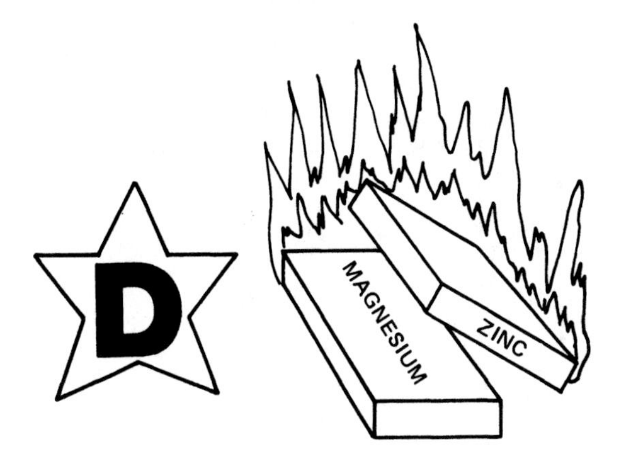

Fig. 3-4 The symbol for type D extinguishers is a yellow star with the letter D in the center. This type of extinguisher is used on flammable metals.

TRASH

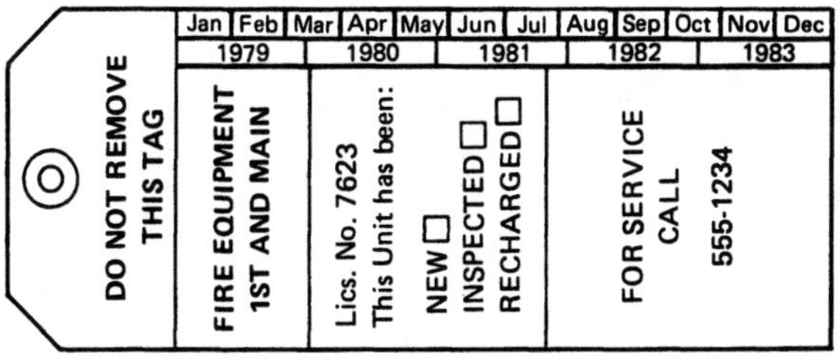

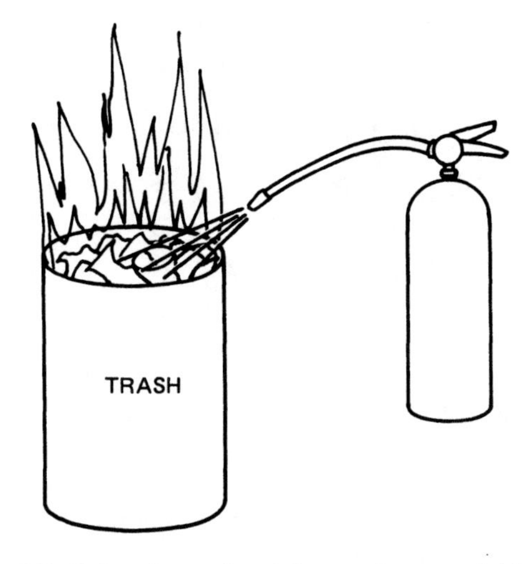

Fig. 3-5 Point the extinguisher at the material burning, not at the flames themselves.

2.2 The use of the wrong type of extinguisher on a fire may be dangerous.

2.3 Point the extinguisher at the base of the fire, figure 3-5.

2.4 Check all extinguishers to see that they have been currently inspected, figure 3-6.

2.5 If the seal is broken, have it reinspected and recharged if necessary.

2.6 Know where all fire extinguishers are located.

Jan Feb Mar Apr May Jun Jul Aug Sep Oct Nov Dec
1979 1980 1981 1982 1983
DO NOT REMOVE
THIS TAG
FIRE EQUIPMENT
1ST AND MAIN
Lics. No. 7623
This Unit has been:
NEW
INSPECTED
RECHARGED
FOR SERVICE
CALL
555-1234

Fig. 3-6 A tag like this should be on all fire extinguishers in the shop. Be sure the date marked on it has not passed.

3. LOCATING AND MARKING FOR FIRE EXTINGUISHERS

3.1 The type of extinguisher should cover the type of combustible materials found in the area, figure 3-7.

3.2 An extinguisher should be located just inside the door so it can be reached easily.

3.3 Locate an extinguisher near flammable materials, but not so close that it cannot be reached in a fire.

AUTO SHOP
OIL
WOOD
WALL
TYPE B
TYPE A
ELECTRICAL
BOX
TYPE C
WOOD SHOP

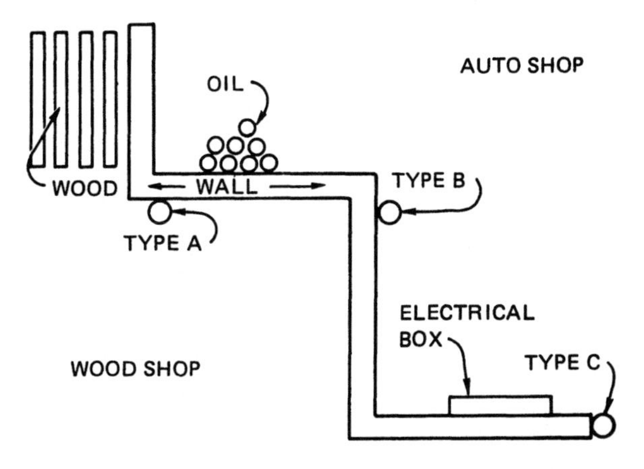

Fig. 3-7 The type of fire extinguisher used should be the kind that will extinguish the type of flammable material in the area.

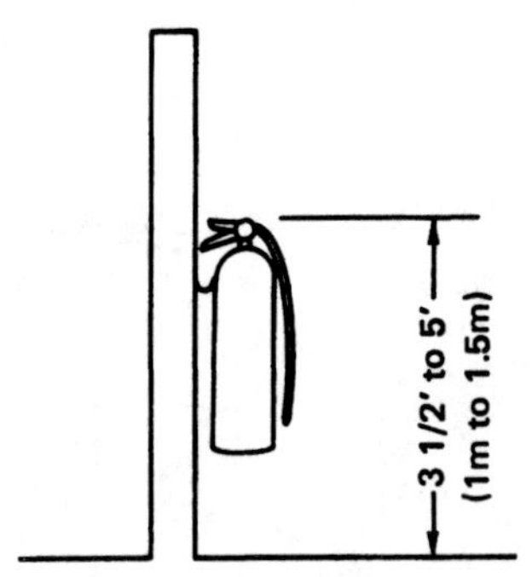

Fig. 3-8 The fire extinguisher should be mounted at the correct height so it can be lifted off easily.

3.4 The top of an extinguisher should be mounted so it is from 3 1/2 feet (1 m) to 5 feet (1.5 m) from the floor, figure 3-8.

3.5 Mark the location of a fire extinguisher so that it can be seen from a distance, figure 3-9.

3.6 Make the marking for an extinguisher on a post or column on all sides, so it can be seen from all areas.

3.7 A marking for an extinguisher should also be near the floor, so it can be seen if the room is full of smoke.

3.8 Locate extinguishers so that there are no obstructions to reaching them.

4. WELDING NEAR FLAMMABLE MATERIALS

It is not always possible to remove flammable materials from the welding area. If they cannot be removed, a person should stand fire watch.

4.1 A fire watch is needed if combustible materials are within 35 feet (10.7 m) of any welding or cutting.

4.2 A fire watch is needed if sparks can reach combustible materials further than 35 feet (10.7 m) away.

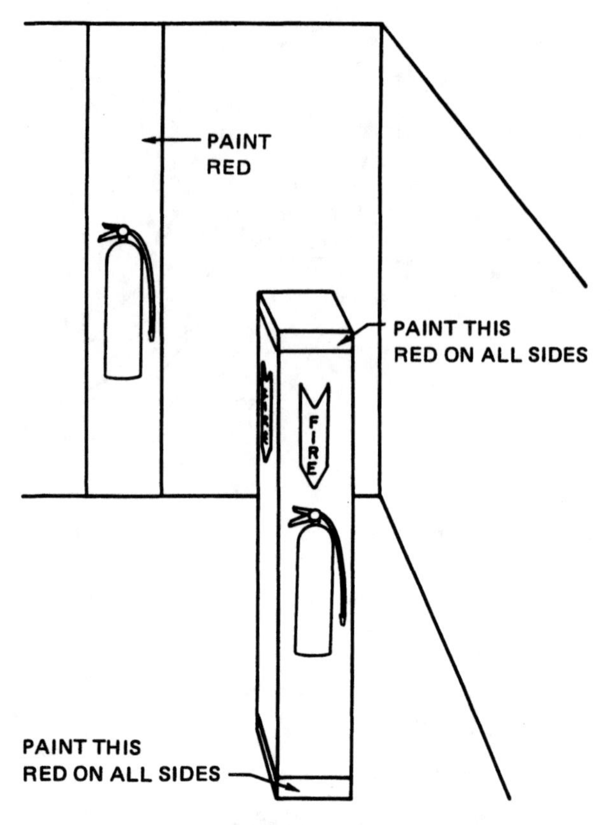

Fig. 3-9 The markings that show locations of fire extinguishers should be as large and easily seen as possible.

4.3 A fire watch is needed on lower levels if sparks can fall through holes in the floor and reach combustible materials on other floors.

4.4 The person standing fire watch must know how to sound the alarm.

4.5 The person standing fire watch must be instructed in the use of the available fire extinguisher.

4.6 Combustible materials that cannot be removed should be covered with wet sand or other fire resistant material before welding.

4.7 Advise a supervisor or instructor and get permission before welding near combustible materials that cannot be removed.

REVIEW QUESTIONS

Read each question and indicate the number of the rule in the unit which most accurately answers the question.

1. How far from the floor should the top of a fire extinguisher be mounted?
2. Where should you point a fire extinguisher when putting out a fire?
3. A fire watch is needed if combustible materials are within how many feet of any welding or cutting?
4. What should you do if the seal is broken on a fire extinguisher?
5. What determines the type of extinguisher to be used?

Read each question and indicate the letter next to the statement that most accurately answers the question.

1. Why should a locating mark for a fire extinguisher be near the floor?
   a. so it can be seen easily when standing close
   b. so it can be seen in a smoke-filled room
   c. because the top may be covered by equipment
   d. all of the above
2. Which type of fire extinguisher would be needed in an area used for storing paint and paint thinner?
   a. type A
   b. type B
   c. type C
   d. type D
3. Which of the following should the fire watch know?
   a. how to sound the alarm
   b. how to use the fire extinguisher
   c. where to find the supervisor or instructor
   d. both a & b
4. When starting work in a new area, you should check which of the following about the fire extinguishers?
   a. their locations
   b. their inspection dates
   c. their operating instructions
   d. all of the above
5. What precaution is needed if sparks may fall down an elevator shaft on combustible materials below?
   a. wet sand to throw down in case of a fire below
   b. a fire hose
   c. a fire watch on each floor below
   d. none of the above

Unit 4 Equipment and Material Handling

OBJECTIVES

After completing this unit, the student will be able to explain the proper method of lifting and moving bulky or heavy objects by hand. The student will also be able to explain the proper safety techniques for using an overhead crane or hoist and for operating a forklift.

The welder can expect to move things that are large, heavy or bulky. As a unit is assembled, its weight and size increases with each part added. Caution needs to be taken not to overload the people or equipment moving the assembly.

1. LIFTING AND MOVING BY HAND

1.1 Lift using your leg muscles, not your back, figure 4-1.

1.2 Distribute the weight equally between both hands.

1.3 Get help lifting if the object weighs more than you could lift with one hand.

1.4 Get help if the object is bulky or awkward.

1.5 Have two or more people carry long objects, one at each end.

1.6 When carrying things, walk with your back straight.

1.7 One person should have the responsibility for giving directions before a weight is lifted and carried with others.

1.8 Avoid lifting over your head.

1.9 If it is necessary to lift overhead, wear a hard hat and eye protection to prevent injury from any dirt or any unseen material that may be on top of the object you are moving, figure 4-2.

1.10 Get a taller ladder to avoid standing at the top of a shorter one.

Fig. 4-1 Your back should be straight so that you lift with your legs.

Fig. 4-2 You never know what may be on top of the box you are lifting.

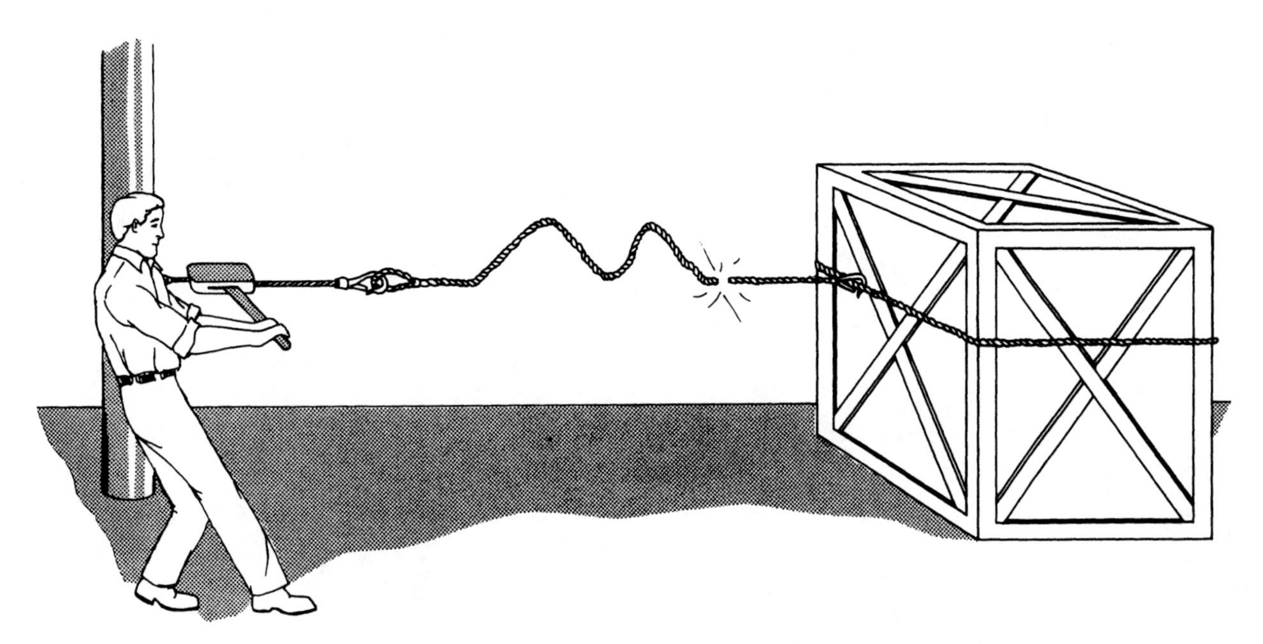

Fig. 4-3 If you are standing to one side when a rope breaks, it will miss you. It can snap back like a broken rubber band.

1.11 When moving a load, use a stronger come-along or winch rather than a weaker one with a cheater (a piece of pipe) for leverage.

1.12 Stand to one side when using a come-along, chain, cable, or rope to pull, so that if something breaks and snaps back it will miss you, figure 4-3.

2. LIFTING AND MOVING WITH A CRANE OR HOIST

2.1 Stand clear of anything being lifted overhead.

2.2 Make sure all chains and clamps are tight before lifting.

2.3 Be sure the crane or hoist has a load rating large enough to lift the load safely.

2.4 For lifting, chains are better than cables or ropes.

2.5 Check the load rating if plate clamps are used, figure 4-4.

2.6 Some types of self-locking plate clamps may release the load if tension is not kept on the clamp at all times.

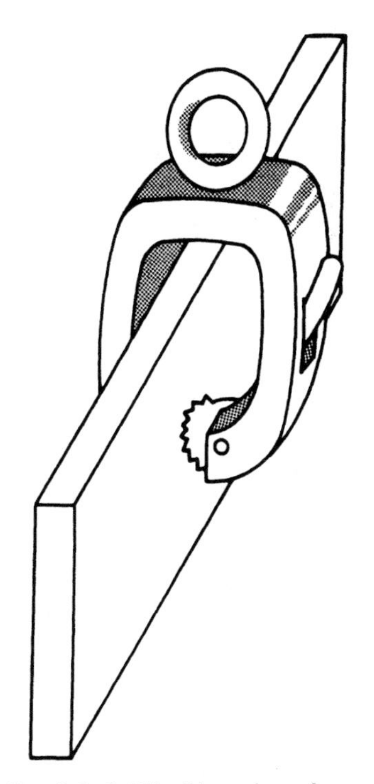

Fig. 4-4 Self-locking plate clamp

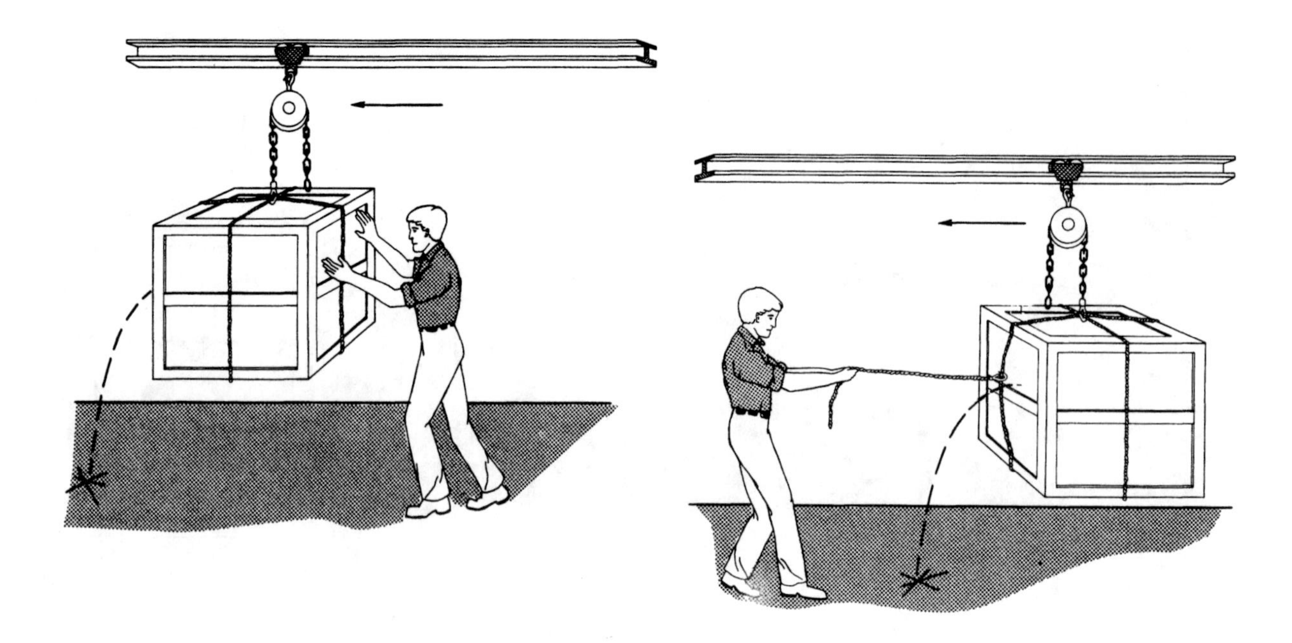

Fig. 4-5 If the load falls while being moved, it will miss you.

2.7 Pushing a load on a hoist is better. If it must be pulled, use a rope, figure 4-5.

2.8 If a load is to be moved over a work area, any workers in the area should move.

2.9 Check with your supervisor or instructor before using any crane or hoist for the first time.

3. USING A FORKLIFT

3.1 Have the forks spread apart as wide as possible under the load.

3.2 The forklifts should be about 3 inches (75 mm) from the edge of the load unless there are special cutouts for them.

3.3 Lift and carry the load all the way back on the forks, figure 4-6.

3.4 Carry the load so that the weight is centered on the forks.

3.5 Know the lifting capacity of the forklift and never overload it.

Fig. 4-6 The load should be all the way back on the forks and no more than 1 inch above the ground when being moved.

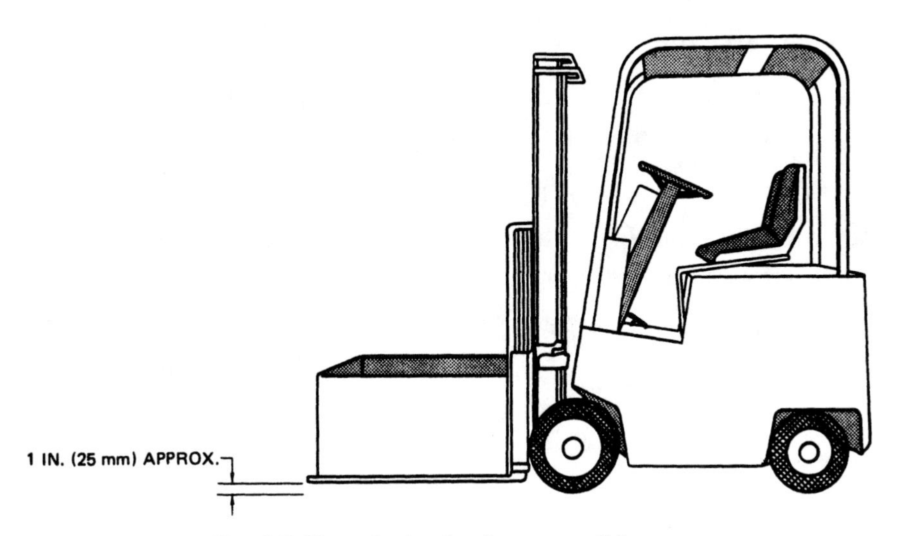

Fig. 4-7 Keep the load as low as possible.

3.6 The load should be as low as possible while it is being moved, figure 4-7.

3.7 Drive up and back down ramps with a loaded forklift to prevent the forklift from tipping over, figure 4-8.

3.8 The fork truck must be level when lifting any load.

3.9 Watch the wheels when moving a load. If they drop in a hole or run over something, the load may shift or fall.

3.10 Never use a forklift to carry or lift people.

3.11 Check with your supervisor or instructor before using any forklift for the first time.

Fig. 4-8 The load on a forklift should be carried on the uphill side when going up or down a ramp.

REVIEW QUESTIONS

Read each question and indicate the number of the rule in the unit which most accurately answers the question.

1. What protection should you have before you lift over your head?
2. What should you do if you plan to move a load over an area where there are workers?
3. How high should a load be carried on a forklift?
4. Where should you stand when using a chain, cable, or rope with a come-along to move something?
5. What can happen if the tension is taken off some self-locking plate clamps?

Read each question and indicate the letter next to the statement that most accurately answers the question.

1. When is it safest to be lifted by a forklift?

   a. if you have a guardrail
   b. if you are lifted less than 10 feet (3 m)
   c. if you are lifted for less than 5 minutes
   d. you should never be lifted

2. How heavy an object should you try to lift by yourself?

   a. no heavier than you can lift with one hand
   b. less than 50 pounds (22.7 kg)
   c. no more than you can lift without your back hurting
   d. less than 25 pounds (11.3 kg)

3. In which direction should you drive down a ramp with a loaded forklift?

   a. with light loads, drive down forward
   b. always drive down forward so you can see better
   c. always back down ramps
   d. none of the above

4. Which of the following ways is the best to move a load on an overhead hoist?

   a. pull it toward you
   b. pull it with a rope
   c. push it away from you
   d. both b and c

5. Who should you check with before operating a forklift for the first time?

   a. a friend
   b. someone who has operated it before
   c. your supervisor or instructor
   d. no one, if you think you know how

Unit 5 Tool and Equipment Usage

OBJECTIVES

After completing this unit, the student will be able to explain how to safely use power equipment and hand tools.

1. HAND TOOLS

Occasionally the welder is required to use hand tools to disassemble or assemble parts for welding, as well as to perform maintenance and repairs on equipment.

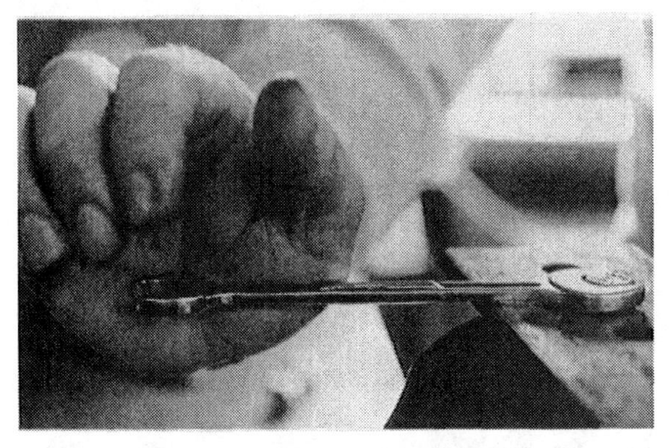

Fig. 5-1 Pushing a wrench like this will help avoid skinned knuckles if the wrench slips.

1.1 When turning a tight bolt or nut with a wrench, push the wrench with an open hand, figure 5-1.

1.2 Get a bigger wrench with a longer handle if the wrench handle is too short to get enough leverage to turn the nut or bolt.

1.3 More force can be applied to a breaker bar than can be applied to a ratchet handle without damaging the wrench.

1.4 The fewer the number of points (number of points means the number of different positions a wrench will fit on a nut), the more pressure can be applied without damaging the nut or bolt head, figure 5-2.

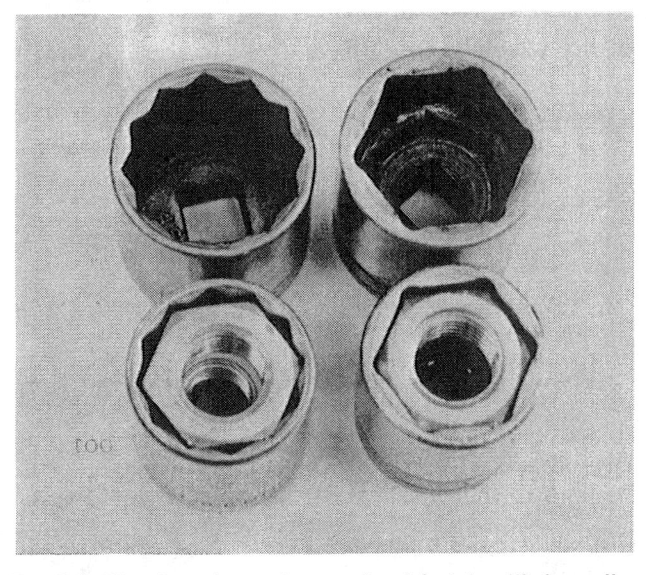

Fig. 5-2 The six-point socket on the right is less likely to slip or damage the nut than the twelve-point socket on the left.

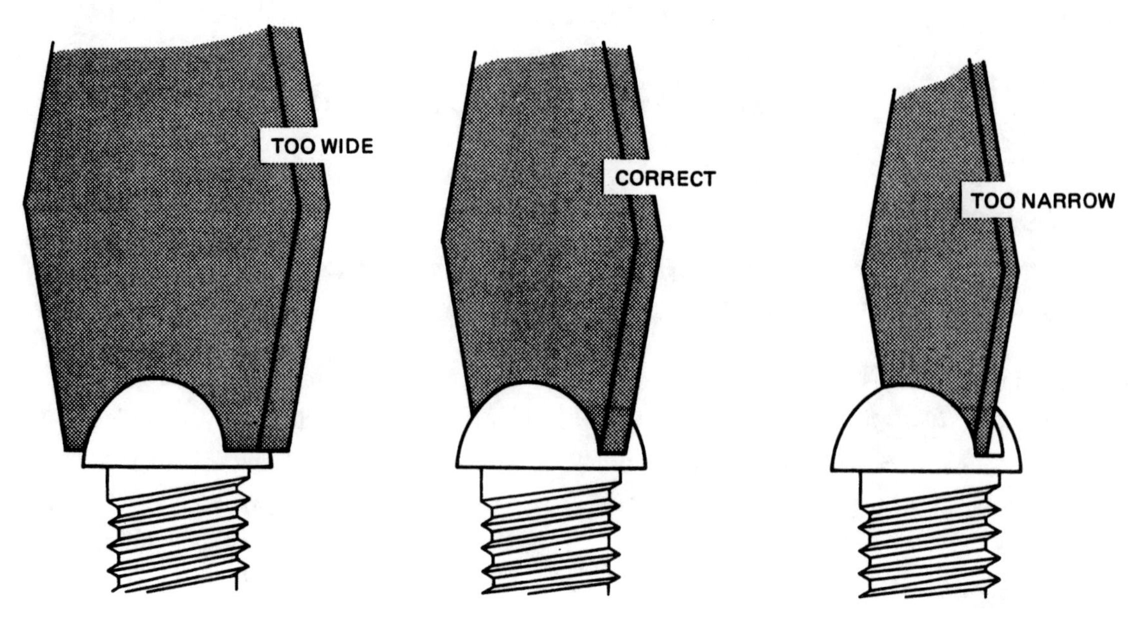

Fig. 5-3 Choosing the correct screwdriver size for the slot is important.

1.5 Box-end wrenches are safer and stronger than open-end wrenches.

1.6 If the wrench is too large and fits the nut or bolt head loosely, it may slip suddenly, causing injury.

1.7 If heat is used to loosen a bolt or nut, the wrench will also get hot. Use caution in handling the wrench.

1.8 The end of the screwdriver should fit the screw head snugly and fill the slot or other head design completely without hanging over the edges, figure 5-3.

1.9 Files are too brittle to be used as pry bars.

1.10 Do not use a screwdriver as a pry bar. The tip and handle may be damaged.

1.11 When using power screwdrivers, use phillips head screws to prevent the tip from sliding out of the slot.

1.12 The heads of all chisels and punches should be ground to remove any mushrooming, figure 5-4.

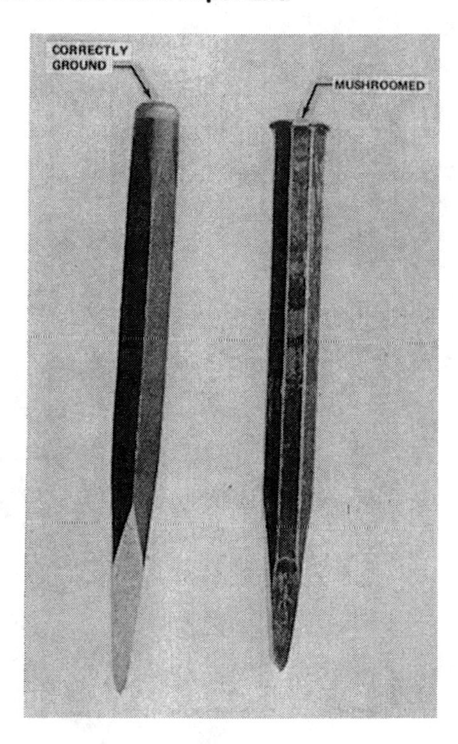

Fig. 5-4 Notice that the correctly ground chisel head has a chamfered edge. This helps prevent future mushrooming.

1.13 Chisels and punches should be kept sharp. A sharp chisel takes less force to do the job and is less likely to glance off.

1.14 When other than light taps are being used on a chisel, use a chisel holder or pliers to eliminate the danger of striking your hand.

1.15 Wear safety glasses when hammering.

1.16 The hammer head should be on the handle securely.

1.17 Replace cracked or damaged hammer handles.

1.18 Light swings are more accurate and less likely to glance off than hard swings.

1.19 Remove all chips or mushrooming on the hammer.

1.20 Everyone in the area should stand well away from the person swinging the hammer so that they will not be struck if the hammer head or hammer flies loose.

2. GRINDING

2.1 Wear safety glasses in addition to using the safety screen or glass provided with the grinder.

2.2 Grinding stones have a maximum rpm (revolutions per minute) listed on the side blotter (paper) and must never be put on a machine that has a higher rpm, figure 5-5.

2.3 Most grinding stones are for grinding ferrous metals such as steel, iron, stainless steel, etc. They will become glazed (the surface clogged with metal), if nonferrous metals such as aluminum, copper, and brass are ground.

2.4 Most grinding stones are designed to be used on the face and not the side. Using the side can make the stone explode. Using the side of a grinding stone makes the stone thinner. The side cannot be refinished, and may become out of balance.

2.5 Grinding stones that have become glazed, dull, uneven or out of balance, must be refinished or "trued," figure 5-6.

2.6 Run the stone at full speed when grinding. Do not slow it down or stop it by pressing too hard on the metal being ground.

2.7 Always use the tool rest to support the metal being ground.

2.8 Adjust the tool rest so that it has approximately a 1/16-inch (1.5 mm) gap between it and the stone, figure 5-7, page 24.

Fig. 5-5 This is a general-purpose grinding wheel that must never be operated at a speed higher than 3600 RPM.

Fig. 5-6 Tools used for truing a grinding stone

2.9 All safety guards must be in place when grinding.

2.10 Use pliers to hold small or hot pieces of metal when grinding. Gloves can be caught by the stone, resulting in severe injury.

2.11 Test a new or used stone for soundness before installing by placing it on a table and lightly tapping the side. A sharp ringing should be heard. Roll the stone and repeat this all the way around the stone on both sides. If a dull sound is heard, the stone may have an internal crack and should not be used, figure 5-8.

2.12 The hole in the stone should be the same size as the grinder shaft.

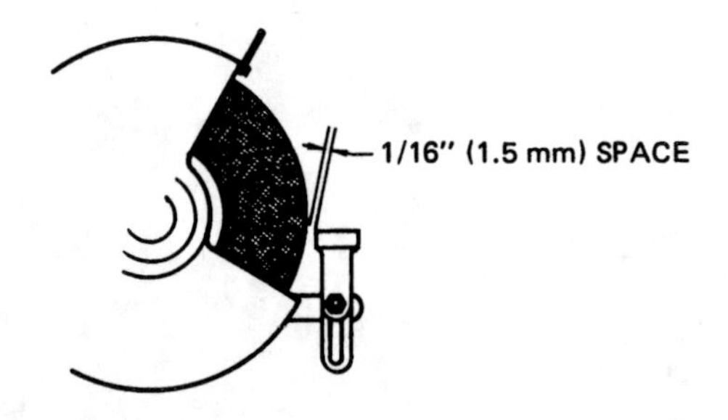

Fig. 5-7 Space between the tool rest and the grinding stone should be no more than 1/16 inch (1.5 mm).

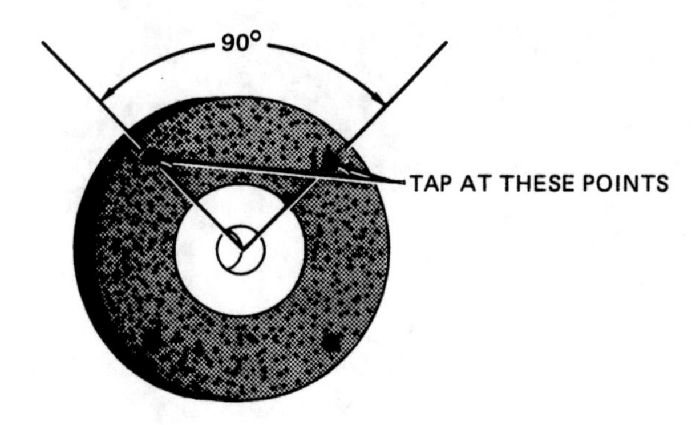

Fig. 5-8 Tap the side of the grinding stone at four points, approximately 90° apart. The stone should be on a hard surface and tapped near the top.

2.13 Stand to one side when starting a grinding stone for the first time.

2.14 The safety guard must be in place when using a portable hand grinder with a cup stone.

2.15 Watch to see that the grinder sparks are not falling on other workers, welding equipment, or any other thing that may be damaged by the sparks.

3. PORTABLE POWER TOOLS

3.1 Wear safety glasses when using any power tools.

3.2 Ground all power tools.

3.3 Extension cords must be 3 wire for grounding and should be as short as possible for the job.

3.4 The wire in an extension cord must be large enough to handle the current requirements of the tool, table 5-1.

Name-plate	CORD LENGTH IN FEET							
Amperes	25	50	75	100	125	150	175	200
1	16	16	16	16	16	16	16	16
2	16	16	16	16	16	16	16	16
3	16	16	16	16	16	16	14	14
4	16	16	16	16	16	14	14	12
5	16	16	16	16	14	14	12	12
6	16	16	16	14	14	12	12	12
7	16	16	14	14	12	12	12	10
8	14	14	14	14	12	12	10	10
9	14	14	14	12	12	10	10	10
10	14	14	14	12	12	10	10	10
11	12	12	12	12	10	10	10	8
12	12	12	12	12	10	10	8	8

Table 5-1 Recommended extension cord sizes for use with portable electric tools. Wire sizes shown are A.W.G. (American Wire Gauge).

3.5 If you feel any electrical shock, stop using the power tool.

3.6 Use power tools only in dry areas.

3.7 If a power tool becomes hot, stop and allow it to cool before continuing work.

3.8 You should have a good footing to prevent a loss of balance if the power tool suddenly catches or jumps.

3.9 Have a good solid hold on the tool before it is started.

3.10 Allow the tool to come to a complete stop before putting it down.

3.11 If the tool becomes jammed, stop immediately.

3.12 Apply steady, even pressure and keep the tool going straight.

REVIEW QUESTIONS

Read each question and indicate the number of the rule in the unit which most accurately answers the question.

1. What should be used to hold a chisel?
2. Where should people stand when you are hammering?
3. How many wires should an extension cord have?
4. Approximately how close to the grinding stone should the tool rest be adjusted?
5. How can you test a grinding stone for cracks before it is installed for use?

Read each question and indicate the letter next to the statement that most accurately answers the question.

1. With which of the following should you hold small, hot pieces of metal?
   - a. gloves
   - b. rags
   - c. pliers
   - d. none of the above
2. What should be done with the head of a chisel that has mushroomed?
   - a. the mushrooming should be ground off
   - b. nothing, it makes the head bigger and easier to hit
   - c. nothing, it will break off itself
   - d. both b & c
3. Most grinding stones are for which type of metal?
   - a. nonferrous, like aluminum and copper
   - b. nonferrous, like stainless steel and iron
   - c. ferrous, like brass and copper
   - d. ferrous, like steel and iron

4. What safety protection should you wear when using any power tools?
   a. a hard hat
   b. safety glasses
   c. leather gloves
   d. all of the above

5. Why should files not be used as pry bars?
   a. they are too brittle and will break
   b. they are too short
   c. they may roll out from under the part being pried
   d. the file teeth may scratch the metal

Unit 6 Protection Clothing and Equipment

OBJECTIVES

After completing this unit, the student will be able to explain what types of clothing should be worn in a welding shop and describe the special types of safety clothing used for welding or cutting.

It is not always possible to wear specially designed protection all the time in a welding shop. The proper selection of general work clothing should be made to minimize the hazards of the shop.

1. GENERAL CLOTHING

1.1 All clothing should be 100 percent wool or cotton.

1.2 All clothing should be thick enough to stop the ultraviolet light from passing through.

1.3 Wear long-sleeved shirts with buttons on the cuffs.

1.4 Shirt pockets should be removed or have flaps.

1.5 Shirts should have a collar to protect the welder's neck, figure 6-1.

1.6 The shirt should be dark in color to prevent light from being reflected.

1.7 Shirt tails should be long enough to tuck into the welder's pants.

1.8 Pants should be long enough so that the pant legs cover the tops of the boots.

1.9 The legs should be straight, without cuffs. Cuffs can collect sparks.

1.10 All-leather boots are the best footwear.

1.11 Boots should be high topped.

1.12 The top of the toe of the boot should be plain, figure 6-2.

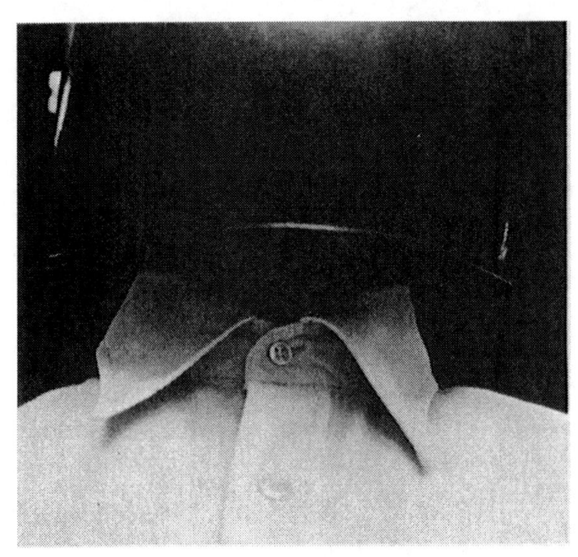

Fig. 6-1 If the top button is left open, the welder's throat could be severely burned.

Fig. 6-2 The top of the boot toe should be smooth so that sparks will not be caught in the seams.

Fig. 6-3 Steel safety-toed boots are required by many welding shops.

1.13 Steel-toed boots are best, figure 6-3.

1.14 Coveralls, lab jackets, or a jump-suit can be worn over most clothing.

1.15 Matches and lighters should not be kept in any pockets. The hot welding sparks may light the matches or burn a hole through a plastic lighter, causing a serious burn.

1.16 Remove or repair all frayed edges on clothing.

1.17 Repair all tears or holes in clothing.

1.18 Folds, creases and large seams can collect sparks.

1.19 A cap can be worn to protect the head.

1.20 Hair long enough to hang down on the chest should be kept in back.

1.21 Wear a hard hat if you are working below others or if material or equipment is moving overhead.

1.22 Wear ear protection around loud noises, such as from power grinders, chippers or hammers, figure 6-4.

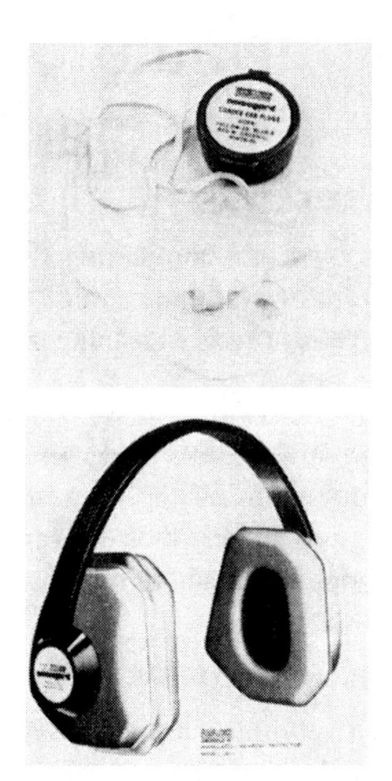

Fig. 6-4 Two types of ear protection.

2. WELDING AND CUTTING PROTECTION

Leather is a fairly good insulator. It resists burning and is flexible enough to be worn with shop clothes for protection.

2.1 Leather aprons are good for protecting your chest and lap when standing or sitting, figure 6-5.

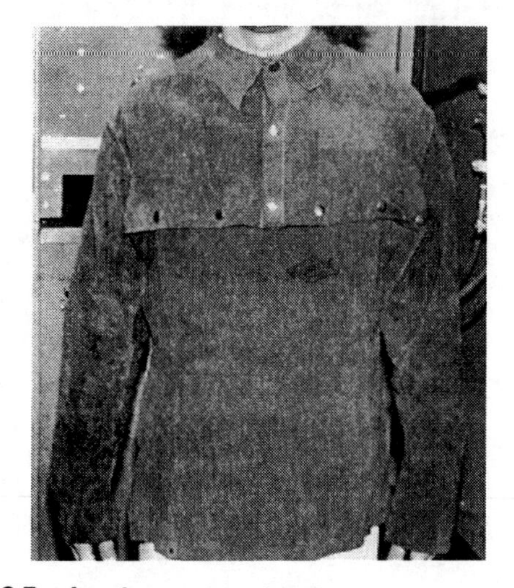

Fig. 6-5 Leather cape with B. B. apron attached.

Fig. 6-6 Full leather jacket

2.2 Leather jackets with full sleeves, back, and a high neck are good for out-of-position work, figure 6-6.

2.3 Leather pants are good for protecting the welder's legs.

2.4 Leather sleeves are available to protect one or both arms, figure 6-7.

2.5 Gloves should be all leather, with or without a lining, and have a gauntlet-type cuff, figure 6-8.

2.6 Leather chaps can be tied on the legs to give protection to the front of the legs.

2.7 Knee pads are good when kneeling to weld on or near the floor.

2.8 Boot protectors can be strapped around the pant legs and boot tops to prevent sparks from bouncing in the top of the boots.

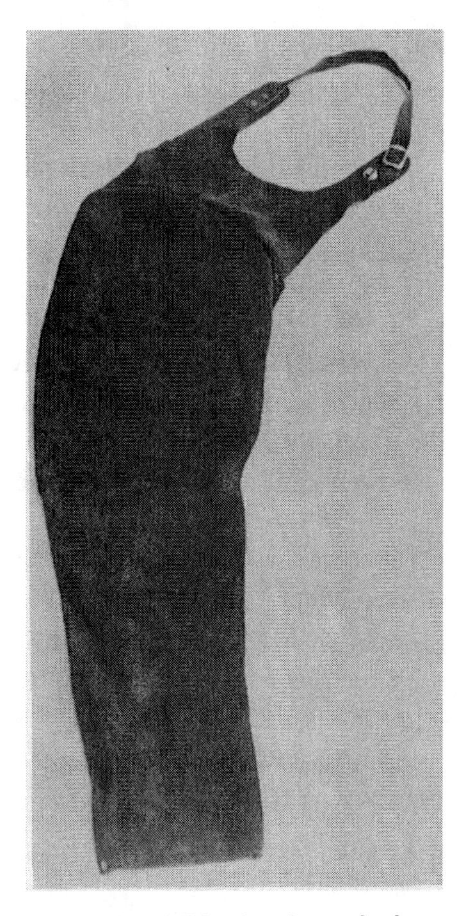

Fig. 6-7 This is a full leather sleeve. It also comes in one-half lengths.

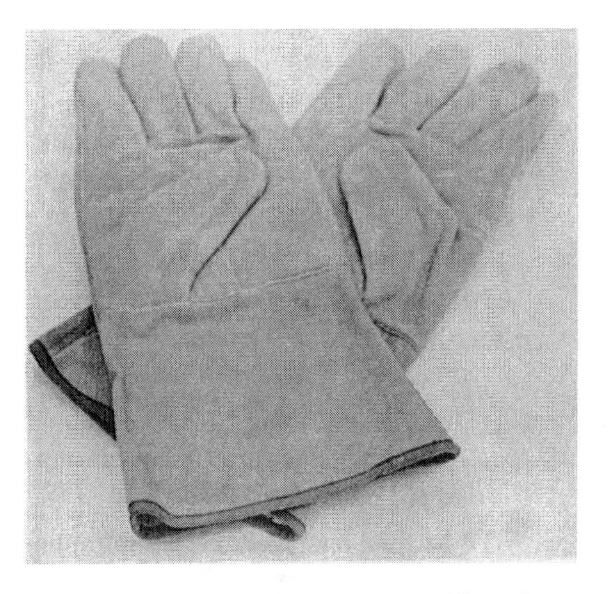

Fig. 6-8 Heavy leather gauntlet welding gloves.

3. FACE AND EYE PROTECTION

3.1 Wear safety glasses with side shields at all times in a welding shop, even under welding helmets and glass goggles.

3.2 Flash glasses (lightly shaded glasses) can be worn as safety glasses. They protect the welder from flying sparks and offer some protection against harmful rays from other work, figure 6-9.

3.3 Never substitute stained glasses, sunglasses, smoked plastic or other materials for any shade (filter) lens for welding.

3.4 Prescription glasses should be plastic or have plastic covers to prevent pitting by sparks.

3.5 Plastic clip-on side shields are available to fit most glasses frames.

3.6 Replace any pitted or cracked shade lens.

3.7 Always use gaskets provided with helmets or goggles, figure 6-10.

3.8 Wear gas welding goggles for all welding, soldering, cutting, and heating.

3.9 Gas welding shade lenses are numbered from light, number 2, to dark, number 8. Most work is done with numbers 4 or 5. Too dark or light a shade will cause eye strain, table 6-1.

3.10 Gas welding goggles are not designed to remove arc welding rays and should only be used for gas work.

3.11 Wear arc welding helmets for all arc welding or cutting operations except submerged arc.

3.12 Arc welding shade lenses are numbered from light, number 9, to dark, number 14. Most work is done with number 10 to number 12. As in gas welding, the correctly numbered shade lens is important, table 6-2.

3.13 The arc welding lens assembly is made up of 3 parts. The outside lens is clear and is either plastic or tempered glass. It protects the shade lens from damage. The center lens is a shade lens. It filters out the harmful light. The inner lens is clear and must be plastic.

Fig. 6-9 Flash glasses

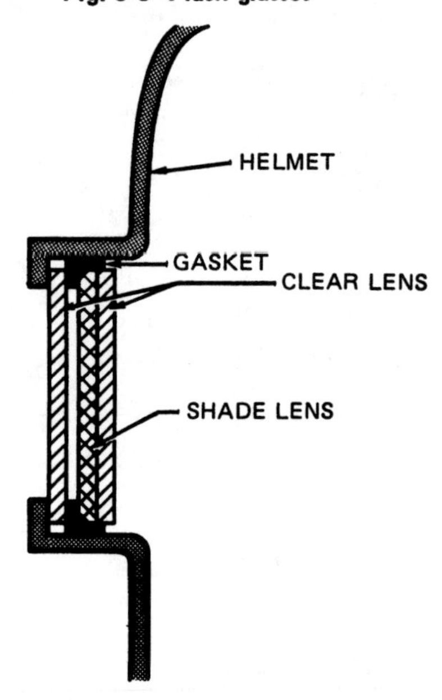

Fig. 6-10 The gasket around the shade lens will stop ultraviolet light from bouncing around the lens assembly.

	Shade Number
Soldering	2
Brazing	3 or 4
Cutting	
Light– up to one inch (25.4 m) thick	3 or 4
Medium– one to six inches (25.4 - 15.24 cm) thick	4 or 5
Heavy– six inches (15.24 cm) and thicker	5 or 6
Welding	
Light– up to 1/8 inch (3.18 mm) thick	4 or 5
Medium– 1/8 (3.18 mm) to 1/2 inch (12.7 mm) thick	5 or 6
Heavy– 1/2 inch (12.7 mm) and thicker	6 or 8

Table 6-1 Gas cutting and welding shade lens numbers

	Shade Number
Shielded Metal-Arc Welding	
Light– 1/16 to 5/32–inch electrodes (1.59 mm - 3.97 mm)	10
Medium– 3/16 to 1/4-inch electrodes (4.76 mm - 6.35 mm)	12
Heavy– 5/16 to 3/8-inch electrodes (7.94 mm- 9.53 mm)	14
Gas Tungsten-Arc Welding	
Nonferrous metals	11
Ferrous metals	12
Gas Metal-Arc Welding	10 to 14
Atomic Hydrogen Welding	10 to 14
Carbon-Arc Welding or Cutting	14

Table 6-2 Electric cutting and welding shade lens numbers

It protects the welder's eyes from the possible breakage of the shade lens, figure 6-13.

3.14 Replace loose or damaged helmets. Dangerous light rays (ultraviolet) are invisible and can get in undetected.

3.15 Clean all face or eye protection used by more than one person between uses.

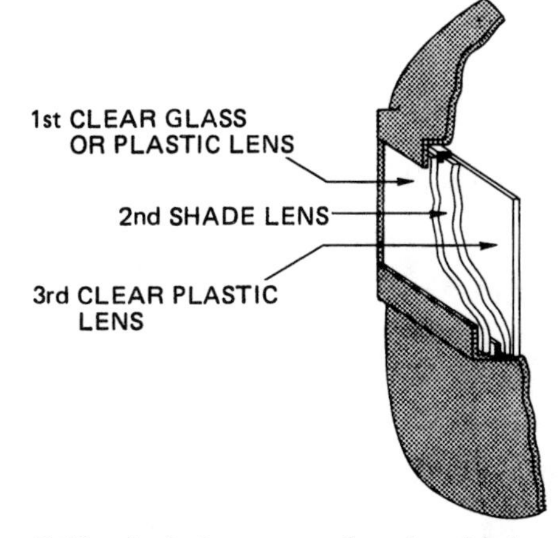

Fig. 6-11 It is important for the third clear lens to be plastic.

REVIEW QUESTIONS

Read each question and indicate the number of the rule in the unit which most accurately answers the question.

1. How long should shirt tails be?
2. Of what materials should all clothing for welding be made?
3. When do you have to wear safety glasses?
4. What should you do with frayed edges on clothing?
5. How long should pant legs be?
6. What protective devices are available for the welder's knees?
7. What type of glove should be worn for welding?
8. Why would you wear flash glases?
9. What is the correct lens assembly for an arc welding helmet?
10. Why should damaged welding helmets be replaced?

Unit 7 Welding, Brazing and Cutting

OBJECTIVES

After completing this unit, the student will be able to assemble and disassemble an oxy-fuel welding and cutting portable cart in a safe and workable manner. The student will also be able to light, adjust and extinguish an oxy-fuel torch in a safe and workable manner.

1. SETUP

1.1 Chain the tanks in the cart before removing safety caps.

1.2 Crack the valve on the tank for a second to blow away dirt that may be in the valve.

1.3 Attach the regulators to the tanks. The nuts should screw on easily and then be tightened with a wrench.

1.4 Never use oil on any oxy-fuel equipment.

1.5 Attach the hoses next. The red hose has a left-handed nut and goes to the fuel gas. The green hose has a right-handed nut and goes to the oxygen.

1.6 Install reverse flow valves at either the torch or regulator end of the hoses. Occasionally, check each reverse flow valve by blowing through it to see if it works properly, figure 7-1.

1.7 Attach the torch next. Install both hose nuts finger tight before using a wrench to tighten either one. Because the threads are very close to each other, the wrench may damage the threads if they are not protected by the other nut, figure 7-2.

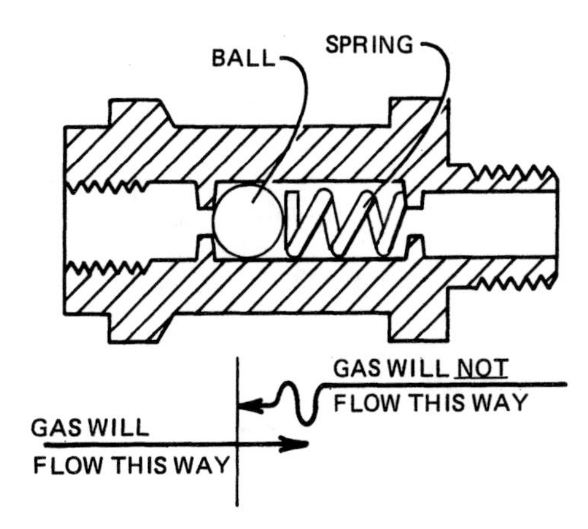

Fig. 7-1 The spring-loaded ball will allow gas to flow one way, but will immediately stop the flow in the other direction.

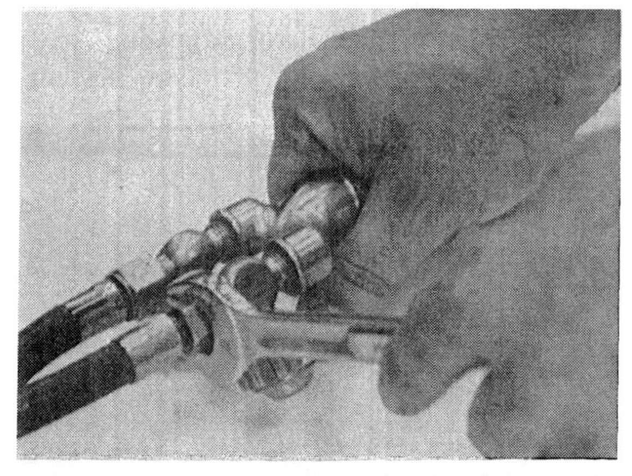

Fig. 7-2 One nut will protect the threads when the other nut is loosened or tightened.

1.8 If the torch is a combination welding and cutting torch, install the attachment next.

1.9 Check the tip seats for necks or O-rings that may be damaged. Some tips are to be hand tightened and others are to be tightened with a wrench. Check in the owners manual or with a supplier for the proper way to tighten your torch. Tightening the wrong way may damage the equipment and be dangerous.

1.10 Back out both regulator adjusting screws until they are loose.

1.11 Open the valve slowly so the pressure rises on the gauge slowly. If the valve is opened quickly, the gauge may be damaged.

1.12 Open the oxygen valve all the way until it is tight at the top.

1.13 Open the acetylene, or fuel valve, 1/4-turn, or enough to get gas pressure. Never open the valve more than 1 1/2 turns, so that in an emergency it can be turned off quickly.

1.14 Open one torch valve next and turn in the regulator adjusting screw until gas can be heard escaping from the torch. The gas should flow long enough to allow the hose to be completely purged (emptied) of air and replaced by the gas before closing the torch valve. Repeat this process with the other gas, figure 7-3.

1.15 When purging the line, be sure there is no flame or spark in the area that might light the escaping gas.

1.16 Adjust both regulators to approximately 5 psi.

1.17 With the torch valve off, spray each connection of the hoses and regulators with a leak detector or soapy water and watch for bubbles which would indicate a leak. Turn off the cylinder valve before tightening any leaky connections.

1.18 To shut off the torch after the flame is out, turn off both cylinder valves.

1.19 Bleed the lines and regulators by opening one of the torch valves at a time and then closing it before opening another valve.

1.20 Back out both regulators' set screws to relieve the stress on the spring and diaphragm.

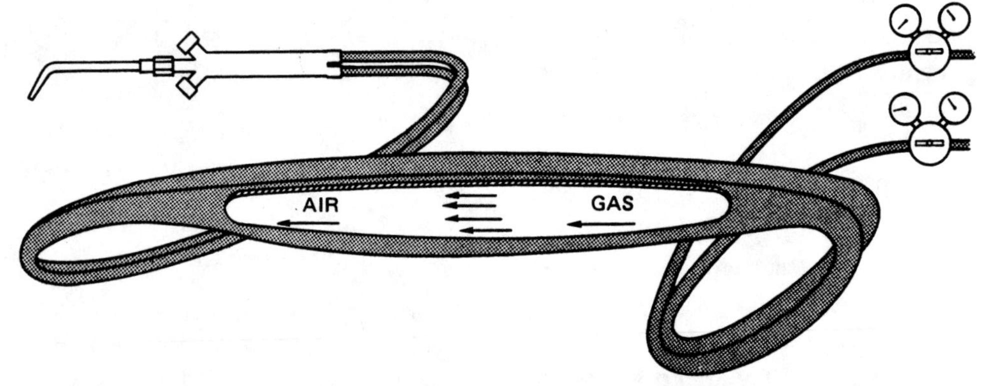

Fig. 7-3 Open the torch valve and the regulator adjusting screw, so gas forces all air out of the hose. The air must be removed because it may contain explosive fumes.

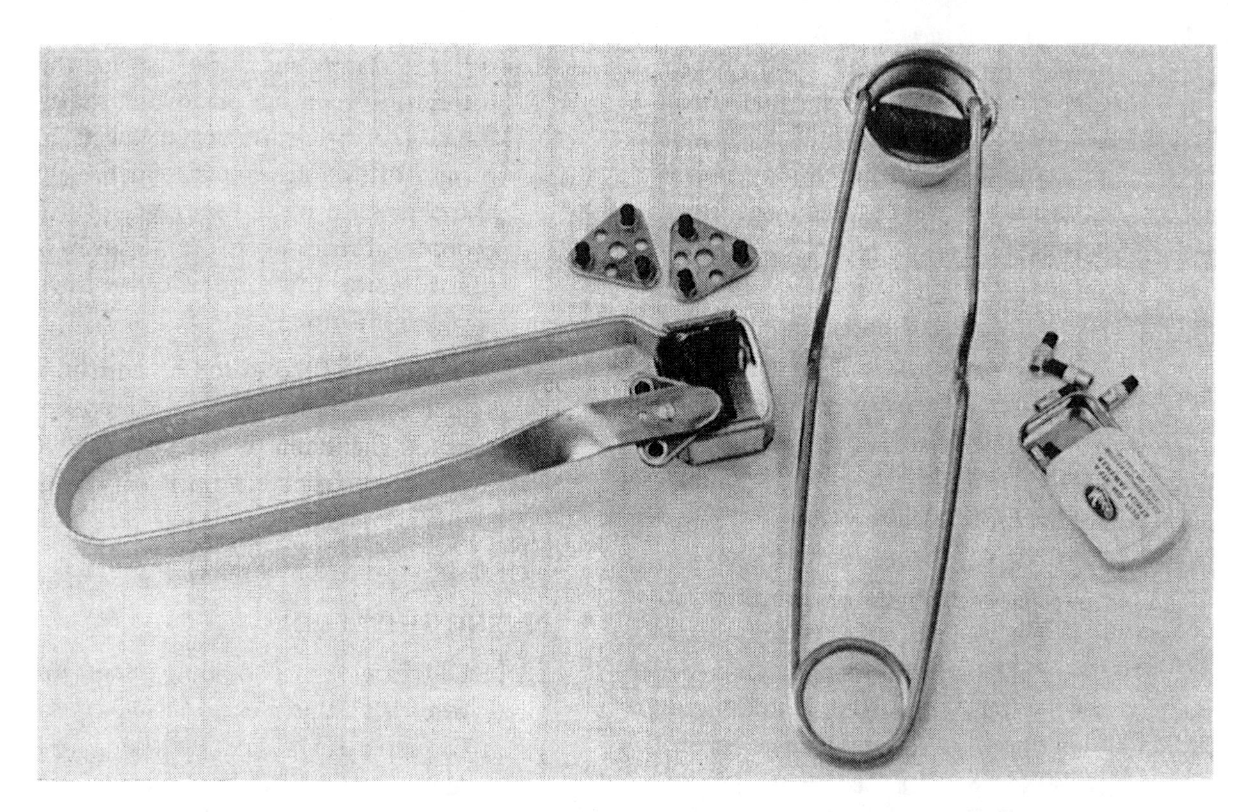

Fig. 7-4 Flint strikers for lighting welding torches. On the left is a single flint. On the right is a more durable flint lighter.

2. LIGHTING AND ADJUSTING AN ACETYLENE FLAME

2.1 Turn on the torch fuel-gas valve just enough to have some gas escape.

2.2 Check to see that the torch is not pointed toward anything that might be flammable or dangerous before lighting it with a spark lighter, figure 7-4.

2.3 A spark lighter is the only safe thing to use when lighting any torch.

2.4 With the torch lit, adjust the acetylene up until the flame stops smoking, figure 7-5.

2.5 Slowly turn on the oxygen and adjust the torch to a neutral flame, figure 7-6.

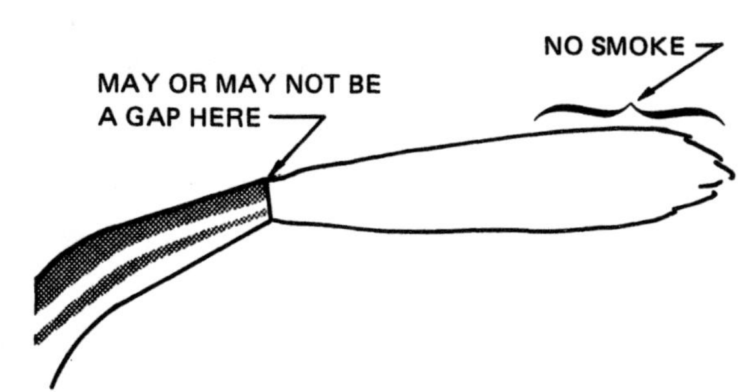

Fig. 7-5 The proper acetylene flow will be smoke-free and help cool the tip. If both the acetylene and oxygen are lit at the same time, the proper acetylene flow cannot be set.

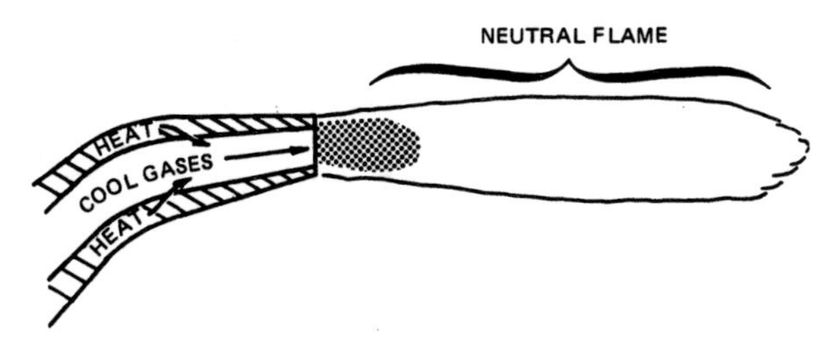

Fig. 7-6 Heat from the tip is pulled out by the cool gases as they pass through the tip.

2.6 When the flame is adjusted properly, but is still too large (hot) or too small (cold) to do the job, change the tip size. The torch may pop if the acetylene is adjusted up or down from the point where it stopped smoking.

2.7 The pressure of both fuel and oxygen may have to be adjusted above 5 psi for longer hoses or larger tips, but should always be as low as possible. The pressure should be equal, except for cutting torches.

2.8 To adjust a cutting-tip flame, set the regulator pressures according to the manufacturer's recommendations and then follow the steps for lighting and adjusting a welding or heating tip. With the flame lit and adjusted, pull the cutting trigger. The flame may become carburizing (without enough oxygen). Readjust it by turning up the oxygen until you have a neutral flame.

2.9 The cutting-oxygen pressure should be as low as possible to do the job. Too high a pressure will throw sparks further and waste oxygen, figure,7-7.

2.10 If the flame burns back inside the torch and does not go out (flashback), or fire starts on a connection, turn off the fuel gas immediately and then turn off the oxygen. The order of turning off the valves is not important as the speed with which it is done.

2.11 To shut off the torch, turn the fuel gas off first and then the oxygen. The flame will be blown out and away from the tip if it is done this way.

3. USING THE WELDING TORCH

3.1 Backfiring, or popping, happens when the flame burns back inside the torch for a second causing it to make a popping sound. The flame immediately reappears or goes out. This is dangerous because it may cause a flashback, or throw hot sparks around the shop.

3.2 Popping may be caused by the tip overheating. This causes carbon in the tip to become hot enough to glow and then the gas mixture in the tip to explode. Popping may

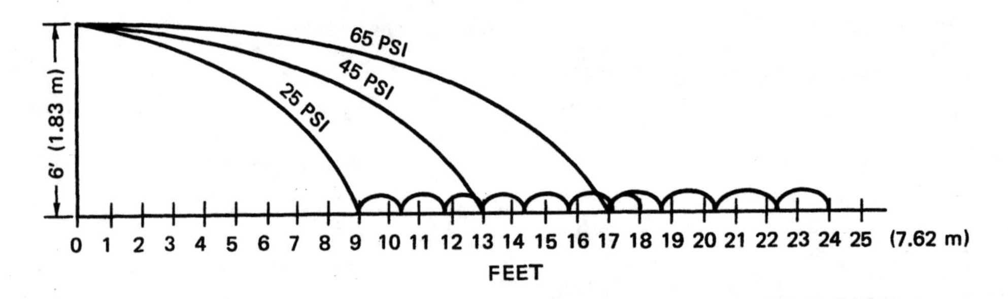

Fig. 7-7 The sparks from cutting a 3/8" (9 mm) thick mild steel plate six feet (1.8 m) from the floor will be thrown much further if the cutting pressure is too high for the plate thickness. This cut was made with a victor cutting tip no. 0-1-101 using 25 psi, as recommended by the manufacturer.

also be caused by gas leaking by one of the torch-tip seats, figure 7-8.

3.3 If the popping is caused by overheating, cleaning the tip and adjusting the flame properly usually will stop the popping. Otherwise, you may be holding the tip too close to the work, holding the torch so that the flame is reflected back on the tip. The torch may also be at the wrong angle to the work, figure 7-9.

3.4 If the popping is caused by a bad tip seat, the tip may have to be replaced and the torch head may have to be reconditioned by the manufacturer.

3.5 To test for a bad tip seat, turn on the oxygen while holding a finger tightly against the end of the tip. Then spray the connecting nut between the tip and torch with a leak-detecting solution. If it bubbles and the nut is tight, it leaks, figure 7-10. The tip seat can also be inspected visually.

3.6 Most fuel gases do not pop as easily as acetylene, and for this and other reasons are safer.

3.7 When cleaning the tip, turn on the oxygen so that it will blow out any dirt loosened during cleaning. Using the flat file provided in the tip cleaner set, file the tip end flat. Next, use the round tip cleaner which easily fits the tip hole to clean the inside of the hole. Caution: Do not use too large a cleaner because it may get stuck or break off in the tip. Also do not push the cleaner in and out repeatedly. It is a file, and cleaning out the hole will make it too large, figure 7-11, page 38.

3.8 When cutting a hole, hold the torch close to the work to heat it up. To start the cut, press the

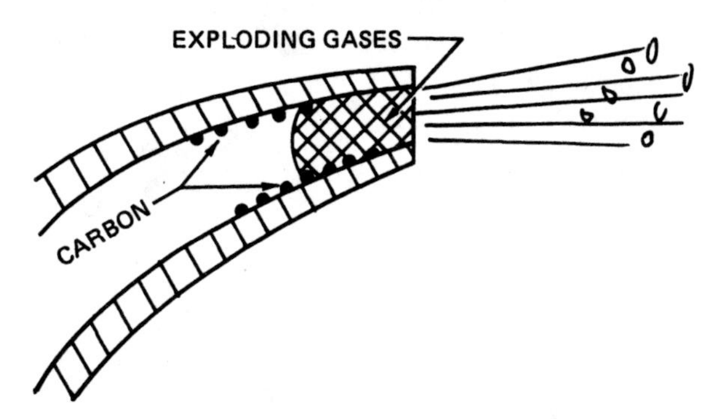

Fig. 7-8 When heat from the flame builds up high enough, the carbon near the end of the tip glows and explodes the gas mixture (popping).

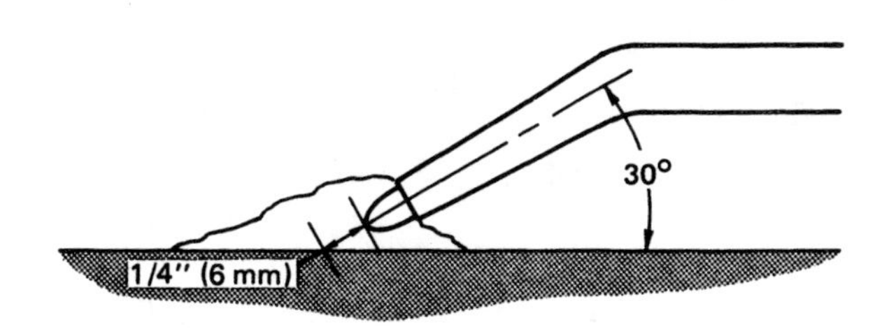

Fig. 7-9 The torch should be held at an angle approximately 30° to the work, and the inner cone should be about 1/4" (6 mm) above the plate.

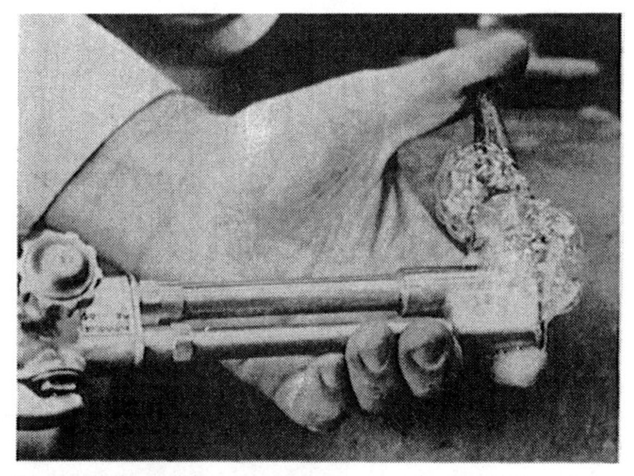

Fig. 7-10 Leaks around the seat cause bubbles when sprayed with a leak detector.

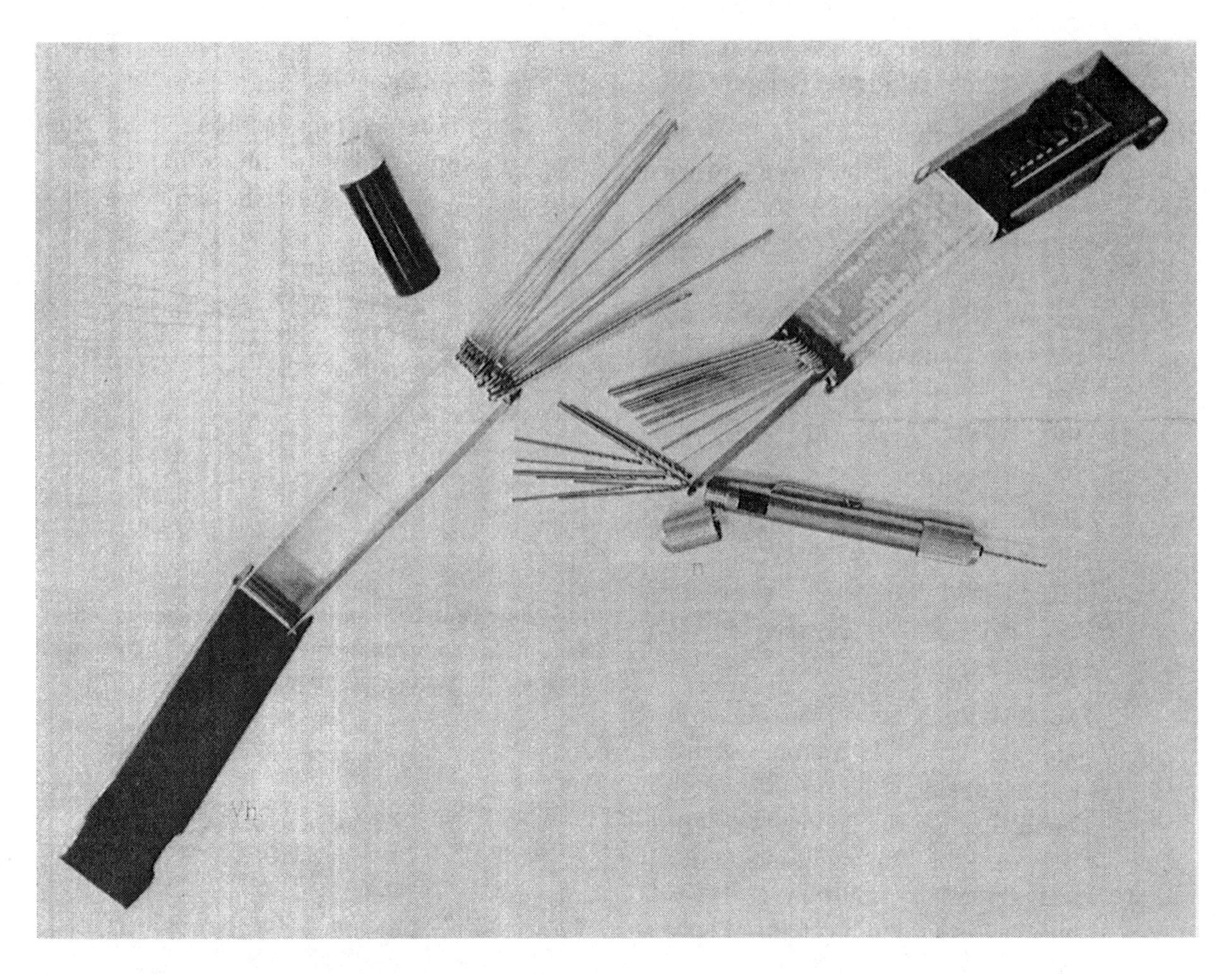

Fig. 7-11 The proper tools are needed to keep the welding tip in a safe and usable condition. The two sets of tip cleaners are for routine maintenance. The set of small drills and the tip-end cutter are used for fixing tips that need major work.

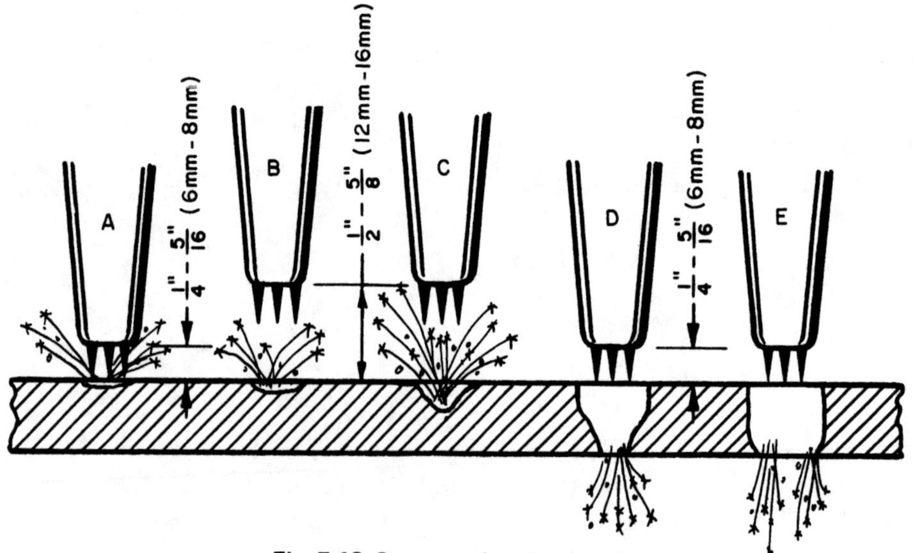

Fig. 7-12 Sequence for piercing plate

cutting trigger slowly and raise the torch up a little as the cut begins. When the hole is completely through the metal, lower the torch and complete the cut as usual, figure 7-12.

3.9 When cutting thick metal, use a cutting torch with a longer handle. This will allow the welder to get further away from the flying sparks.

3.10 Table tops for welding must be made of fire brick or steel plate. Regular brick may pop or explode from the excessive welding heat.

3.11 Torch hoses and regulators should be kept out of the direct line of sparks from a cutting torch.

3.12 Flames, hot metal, and large amounts of hot slag must be kept off of concrete. Rapid vaporization of water in the concrete can cause the concrete to explode.

3.13 To avoid injuries when cutting, be sure the metal is secured so that it will not snap up or fall when the cut is complete.

3.14 When you are leaving any hot metal, write the word "hot" on it in soapstone.

3.15 Before welding or cutting on a used barrel or tank, open or remove one end of the barrel or tank. The tank or barrel should never have contained a flammable liquid.

3.16 There is no totally safe way to weld, braze, solder, or flame cut a used fuel tank.

3.17 When brazing or soldering, keep the heat as low as possible. Most brazing and soldering metals contain lead, zinc, copper or other metals that give off fumes that become hotter. Avoid breathing any fumes rising from the metal or flux.

REVIEW QUESTIONS

Read each question and indicate the number of the rule in the unit which most accurately answers the question.

1. What can cause a torch tip to pop?
2. Why should cutting oxygen pressure be as low as possible?
3. Why is the cylinder valve cracked before the regulator is attached?
4. How do you use a tip cleaner?
5. What is purging?

Read each question and indicate the letter next to the statement that most accurately answers the question.

1. How is a reverse flow valve tested to see if it is working?
   a. see if the torch pops
   b. see if the torch backfires
   c. try to blow back through it
   d. bleed the hose incorrectly and watch the gauges to see if there is any backflow

2. What can happen if a flame or slag is allowed to touch or build up on concrete?

   a. it may explode
   b. it may stick
   c. it may discolor and look bad
   d. it may pit and make it hard to sweep

3. A torch backfire occurs when which of the following happens?

   a. the torch pops
   b. the flame burns back inside the torch and does not go out
   c. when the flame starts hissing
   d. both a & b

4. When lighting an oxyacetylene torch, how high should the acetylene be turned on?

   a. until the flame leaps off the end of the tip
   b. until the flame starts to make a hissing noise
   c. until the flame stops smoking
   d. it depends on the thickness of metal being worked on

5. Before the cylinder valve is opened, the regulator adjusting screw should be:

   a. left alone. It is not moved from the previous pressure adjustment.
   b. it is turned in until there is approximately 5 pounds of pressure when the valve is opened.
   c. turned in 1 1/2 turns for the fuel and all the way for the oxygen.
   d. backed out until it is loose.

Unit 8 Manifold Systems

OBJECTIVE

After completing this unit, the student will be able to identify and list all safety devices that should be on a manifold system.

The use of gas cylinders that are located in a work area can be hazardous. The cylinders are exposed to the moving of equipment, materials and workers. This may cause the cylinders to be knocked over or damaged. A manifold system allows the cylinders to be centrally located and protected from accidental damage.

1. LOCATION AND INSTALLATION OF MANIFOLD SYSTEMS

1.1 Oxygen and fuel-gas manifolds must be 20 feet (6 m) apart or have a 5 foot (1.5 m) high noncombustible wall separating them.

1.2 Fuel-gas manifolds in a work area cannot have more than 300 pounds (136 kg) of liquid or 3000 cubic feet (84 m^3) of gas total capacity connected at one time. Two of these manifolds may be located in the same room if they are separated by 50 feet (15 m) or a noncombustible wall.

1.3 Oxygen manifolds in a work area cannot have more than 6000 cubic feet (168 m^3) of total gas capacity connected at one time. Two of these manifolds may be located in the same room if they are separated by 50 feet (15 m) or a noncombustible wall.

1.4 Fuel-gas manifolds having a total capacity exceeding 300 pounds (136 kg) of liquid or 3000 cubic feet (84 m^3) of gas must be in a separate building or room.

1.5 Oxygen manifolds having a total capacity exceeding 6000 cubic feet (168 m^3) of gas must be located in a noncombustible building or room.

1.6 Special manifold buildings or rooms cannot have open flames and must be well ventilated with lighting fixtures that are gastight and explosion proof, figure 8-1.

1.7 Combustible materials such as paint, oil, and grease cannot be stored within 20 feet (6 m) of an oxygen manifold unless they are separated by a noncombustible wall.

Fig. 8-1 Explosion-proof light fixtures

1.8 Manifolds should be located 20 feet (6 m) or more from stairwells and elevator shafts.

1.9 Inert or nonreactant gas (argon, helium, carbon dioxide, etc.) can be located with the oxygen or fuel-gas manifolds.

1.10 The high-pressure side of a manifold must be equipped with a pressure regulator to provide a working line pressure not to exceed 250 psi.

1.11 Pipe for the high-pressure side of a manifold must be steel, stainless steel, or copper alloys.

1.12 Pipe for the low-pressure side of all manifolds, except acetylene, can be stainless steel, copper (type L or K), brass, steel, or wrought iron.

1.13 All piping for acetylene or acetylenic compounds must be steel or wrought iron.

1.14 Unalloyed copper must not be used for piping acetylene or acetylenic compounds.

1.15 Pipe joints in copper or brass can be welded, brazed with silver alloys over 800 degrees Fahrenheit (427°C), threaded, or flanged.

1.16 Pipe joints in steel or wrought iron can be welded, threaded or flanged.

1.17 All pipe used for manifolds must be clean, oil-free, and should be purged with nitrogen during brazing or welding.

1.18 Install piping at a slight angle so that any moisture in the line will run back toward the manifold, figure 8-2.

1.19 Provide valves for each building so that the gas can be shut off easily in an emergency.

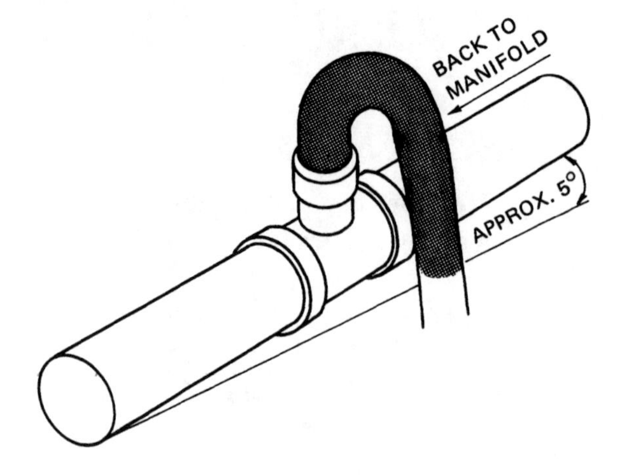

Fig. 8-2 A 5° angle is enough to allow moisture to flow back away from the station regulators. The gas should be taken off the top of the pipe so that returning moisture will not drop down the pipe.

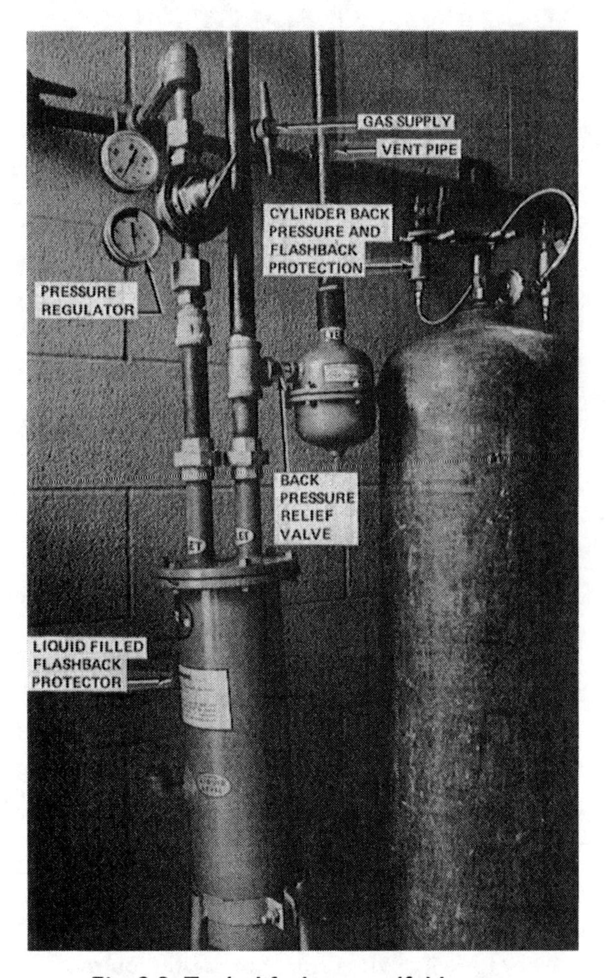

Fig. 8-3 Typical fuel-gas manifold system

1.20 The piping system should be gastight when tested at 1 1/2 times the maximum operating pressure.

1.21 Protective equipment in the fuel-gas line should prevent oxygen from flowing back into the fuel gas, provide flashback protection, and back pressure relief. One device can do all three or separate devices for each are available, figure 8-3.

1.22 Vent back-pressure relief gases directly outside, away from any source of ignition.

1.23 Set back-pressure relief valves at a pressure not to exceed 20 psi.

1.24 Provide each fuel-gas cylinder connected to a manifold with a backflow valve.

1.25 All station outlets must have a shut-off valve on each manifold gas line, figure 8-4.

2. MANIFOLD OPERATION

2.1 Clean out the manifold system, including leads, with an oil-free noncombustible fluid before the regulators are attached. Solutions of caustic soda or trisodium phosphate are good for this purpose.

2.2 After the system is charged, bleed it (allowing gas to escape freely) until the purging gas has been replaced with the gas that is going to be used in the system.

2.3 While bleeding the line, be sure that the escaping gas is well away from any source of ignition.

2.4 Set the line pressure as low as possible to satisfactorily complete the work being done at the work stations.

2.5 Turn off the system when work is completed. Each station and each

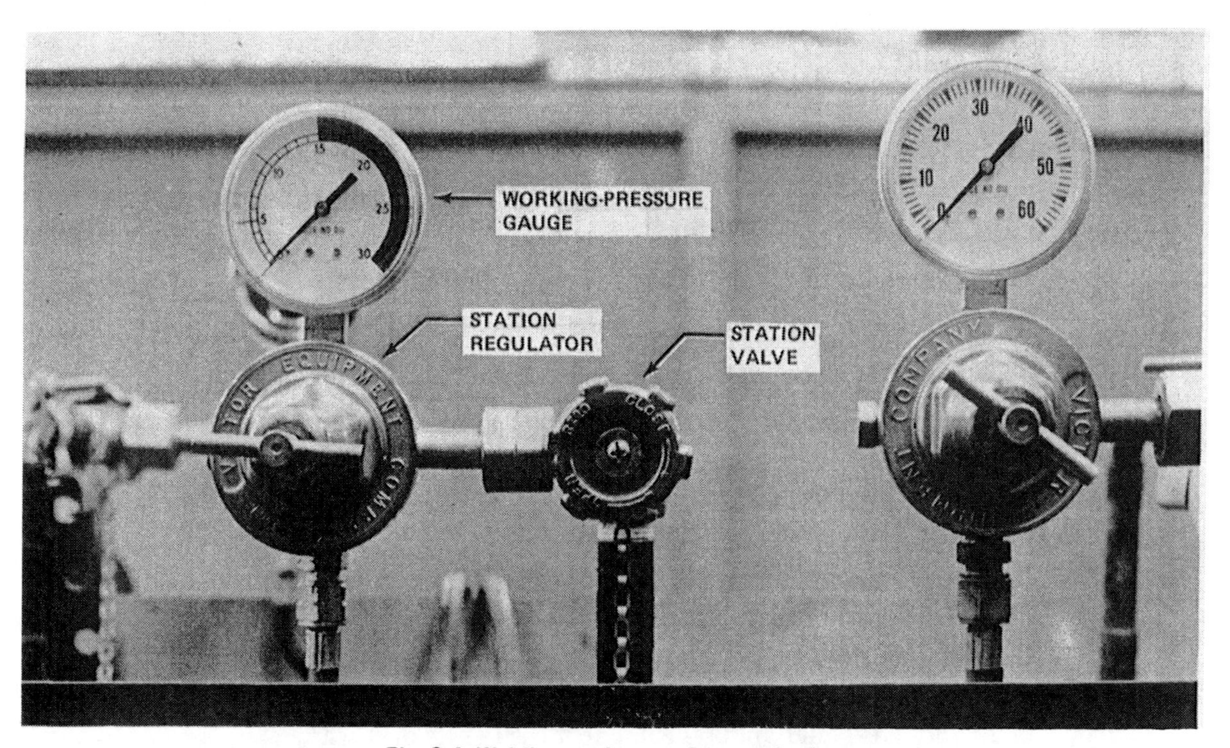

Fig. 8-4 Welding station regulator and valve.

valve on the manifold should be off.

2.6 If a station regulator is removed, put a cap on the line to prevent air from entering the lines, figure 8-5.

2.7 Up to 4 torches can be operated off one station regulator. This does not include machine torches.

Fig. 8-5 Station caps will prevent leaks and keep the valve clean.

2.8 Cylinders attached to a manifold do not have to be chained in place.

2.9 One nonadjustable wrench is needed for each fuel-gas manifold with any cylinders that do not have hand wheels on their valves, figure 8-6.

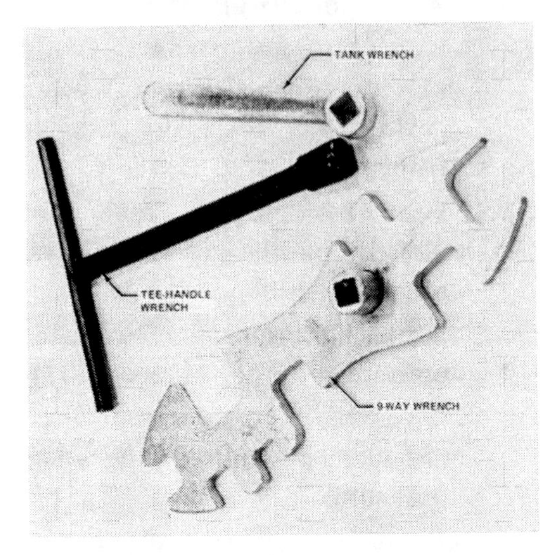

Fig. 8-6 Nonadjustable wrenches for turning on gas cylinders.

REVIEW QUESTIONS

Read each question and indicate the number of the rule in the unit which most accurately answers the question.

1. What solutions can be used to clean the pipes of a manifold system before it is used?
2. How close to stairwells or elevators can a manifold be located?
3. What 3 protections must be provided in a manifold fuel-gas line?
4. What material can be used for piping acetylene or acetylenic compounds?
5. How many torches can be attached to one station regulator?

Read each question and indicate the letter next to the statement that most accurately answers the question.

1. A fuel-gas manifold can have no more than how many pounds of liquid or cubic feet of total gas connected at one time?

   a. 300 pounds liquid or 3000 cubic feet gas (136 kg or 84 m^3)
   b. 300 pounds liquid or 2000 cubic feet gas (136 kg or 56 m^3)
   c. 200 pounds liquid or 3000 cubic feet gas (91 kg or 84 m^3)
   d. 200 pounds liquid or 2000 cubic feet gas (91 kg or 56 m^3)

2. Manifold piping for acetylene must be made of which of the following materials?

   a. steel
   b. type L or K copper
   c. any copper
   d. PVC

3. Manifold line pressure should be set approximately how high?

   a. oxygen 60 psi, fuel gas 15 psi
   b. oxygen 100 psi, fuel gas 12 psi
   c. oxygen 25 psi, fuel gas 7 psi
   d. as low as possible for the job

4. How should steel manifold piping be connected?

   a. welded
   b. threaded
   c. flanged
   d. any one of the above

5. Where can inert-gas manifolds be located?

   a. only with the oxygen manifold
   b. only with the fuel-gas manifold
   c. with either the oxygen or fuel-gas manifolds
   d. only by themselves

Section 3 ELECTRIC

Unit 9 Welding and Cutting

OBJECTIVE

After completing this unit, the student will be able to demonstrate the safe method of using arc welding and cutting equipment.

1. GENERAL ARC WELDING OR CUTTING

1.1 With the welding hood down, the welder is blinded to dangerous things that might be around him. Therefore, it is important that others in the area watch out for any dangerous situations and alert the welder.

1.2 The welder's sense of balance may be affected with the welding hood down, so it is important to have a solid footing and something to hold onto.

1.3 Hold the weight of the welding lead in one hand while welding with the other hand. The welder should not wrap the lead around the arm or body to release the weight. The welder could be pulled off balance and fall if the lead is jerked.

1.4 When welding outside, or in an area without forced ventilation, stand so the welding fumes do not rise directly into your face.

1.5 The welder's face should be more than one foot from the arc when welding or cutting with high amperage settings (300 amps and up). Otherwise, the shade lens may explode from excessive heat.

1.6 Use a safety harness when welding from staging or above one floor level.

1.7 Use a lifeline when entering any area through a small manhole. Attach the lifeline so that the welder can be pulled out quickly in an emergency.

1.8 Keep welding cables out of reach of the welding sparks.

1.9 Welding light may be reflected into the back of the welding hood during welding on aluminum, stainless steel or other reflective metal. Use a properly fitting hood to prevent this.

1.10 Slag on a weld that is cooling may pop free by itself. Slag is usually hot, sharp, and will stick to any exposed skin, causing burns or cuts.

1.11 Chip slag so that the pieces fly away from you, figure 9-1.

2. SHIELDED METAL-ARC WELDING (S.M.A.W.)

2.1 Remove electrodes from electrode holders when leaving the welding area for any time.

2.2 Stand to one side of the falling sparks when welding overhead.

Fig. 9-1 Chip so that the hot slag will fly away from you.

Fig. 9-2 This electrode stub holder can be made out of scrap pipe.

Protective clothing lasts longer if not subjected to direct sparks.

2.3 Burn electrodes down approximately 1 1/2 to 2 inches (38 mm to 50 mm). Burning them shorter damages the electrode holder insulators and may result in accidental shorting out.

2.4 Put electrode stubs in a box. If stubs are thrown on the floor, the welder may slide or fall on them, figure 9-2.

2.5 Do not touch a partially used electrode. It may be very hot after welding.

GAS TUNGSTEN-ARC WELDING (G.T.A.W.)

The high frequency used to start or stabilize the arc can leak through some insulators and cause an electrical shock. This shock is not dangerous, but can be uncomfortable and should be avoided.

3.1 Some G.T.A.W. torches are water cooled. Small water leaks may cause enough dampness to allow a high-frequency shock. Large water leaks can lead to high-voltage shocks from the main power supply of the welding machine. These shocks can be deadly. Shut off and disconnect the welding machine until the water leak is fixed.

3.2 Because of the high degree of proficiency needed for most G.T.A.W. welding, many gloves may be too bulky. Use deerskin or light cotton gloves, figure 9-3, page 48. These gloves provide protection from ultraviolet light rays and high-frequency shock.

3.3 Do not allow the back end of the filler rod to touch a ground during welding. This may conduct the arc current by accident, causing the filler rod to become hot very fast and burn your fingers.

3.4 The tungsten will stay hot for a short time after welding is stopped Be careful not to touch it until it has had time to cool.

3.5 Keep G.T.A.W. leads up so that they are not stepped on or driven over. Damaging them may cause water or gas leaks.

4. GAS METAL-ARC WELDING (G.M.A.W.)

4.1 Because most G.M.A.W. is unshielded and at a high amperage (300 amps and up), unprotected skin will burn quickly.

4.2 Never lay a G.M.A.W. gun down so that the trigger can accidently be pushed, or leave a machine running unattended. If the wire is fed out accidently, it can cause a great deal of damage.

4.3 Never allow G.M.A.W. wire to feed out more than the prescribed distance from the contact tip before touching the ground, figure 9-4. Long pieces of wire will become red hot, and may explode when they are grounded because of the high welding currents and the small size of the wire.

4.4 Remove spatter from the nozzle regularly if it builds up, figure 9-5. It may bridge across the insulator, causing the nozzle to become electrically charged, and it may short out against the metal.

4.5 Additional ventilation may be needed when carbon dioxide gas is used for shielding, figure 9-6. Carbon monoxide, which is poisonous, may be formed by the arc.

4.6 Flux-cored welding may require additional ventilation because of the large amounts of fumes it gives off.

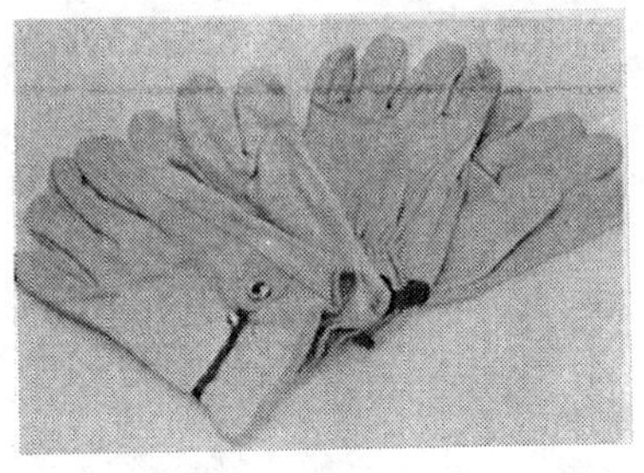

Fig. 9-3 Soft leather or cotton gloves are ideal for G.T.A.W.

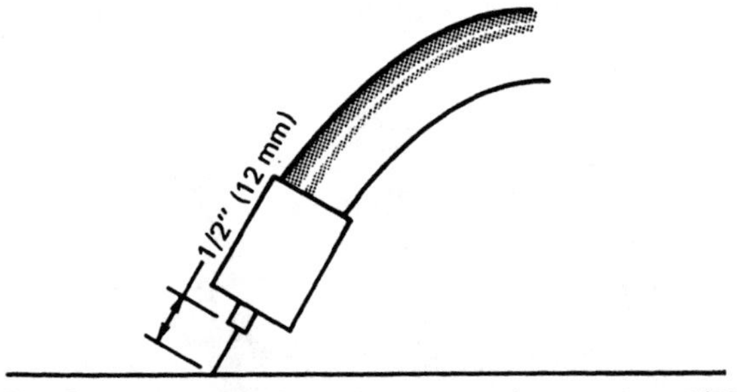

Fig. 9-4 The wire should not extend more than 1/2" (12 mm) beyond the tip.

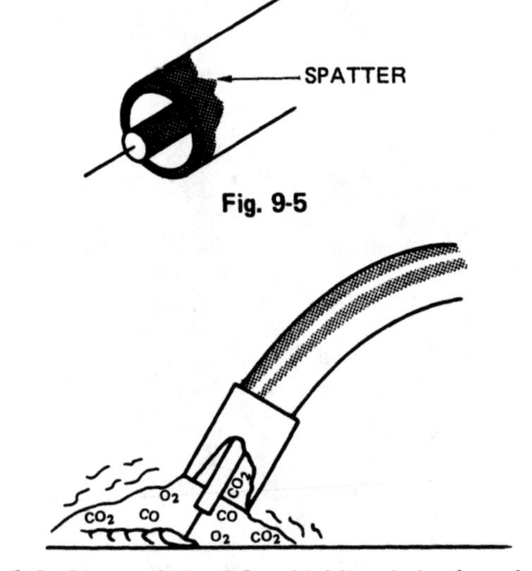

Fig. 9-6 Some of the CO_2 shielding is broken down into CO gas.

4.7 Ozone may be formed by the arc when using shielding gases with oxygen in them, such as carbon dioxide and 95 percent argon/ 5 percent oxygen. Ozone is not dangerous, but can cause a dry throat if a large quantity is inhaled.

4.8 When bare welding wire with a high silicon content is used, glass will form in spots along the weld. It may pop off as the weld cools. The glass will be hot and sharp, so use eye protection.

5. OTHER ELECTRIC WELDING PROCESSES

5.1 Spot welders can throw sparks, so use eye protection.

5.2 If the spot welding presssure is too low to force both pieces of metal tightly together during the welding cycle, the welding current may heat up the metal some distance from the spot weld. This may cause burns.

5.3 Spot welds may have sharp metal sticking up around the edge of the weld. Wear gloves to prevent cuts when handling the welded metal.

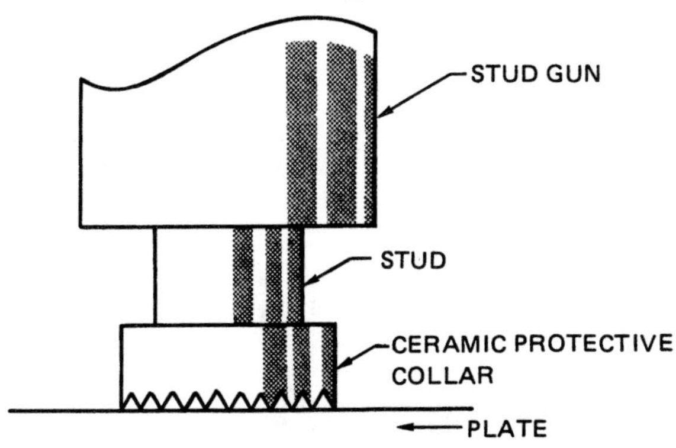

Fig. 9-7 The ceramic collar keeps the sparks and molten metal in place only if it is squarely fitted on the plate.

5.4 Spot welded metal is hot and may cause burns if handled immediately after welding.

5.5 Stud welding can throw sparks if the stud welding gun is not held squarely on the work, figure 9-7.

5.6 Large surges of current in stud welding may cause the welding leads to jump around if the leads are coiled up.

6. AIR CARBON-ARC CUTTING

6.1 Heavy protective clothing is required for air carbon-arc cutting because of the amount of heat and hot sparks.

6.2 Air pressure should be as low as possible for the job so that sparks are not thrown farther than necessary.

6.3 Aim sparks away from other workers and equipment.

6.4 Use ear plugs.

6.5 The wrong polarity will cause the carbon to overheat and break off. This large piece of hot carbon may fly farther than the hot sparks and cause a fire hazard.

6.6 Use extra ventilation when air carbon-arc cutting because of the large volume of fumes.

7. PLASMA ARC CUTTING

7.1 Plasma arc cutting is a noisy process. Use ear plugs.

7.2 Slag from a plasma cut is very hot. Direct it into a slag bin when possible.

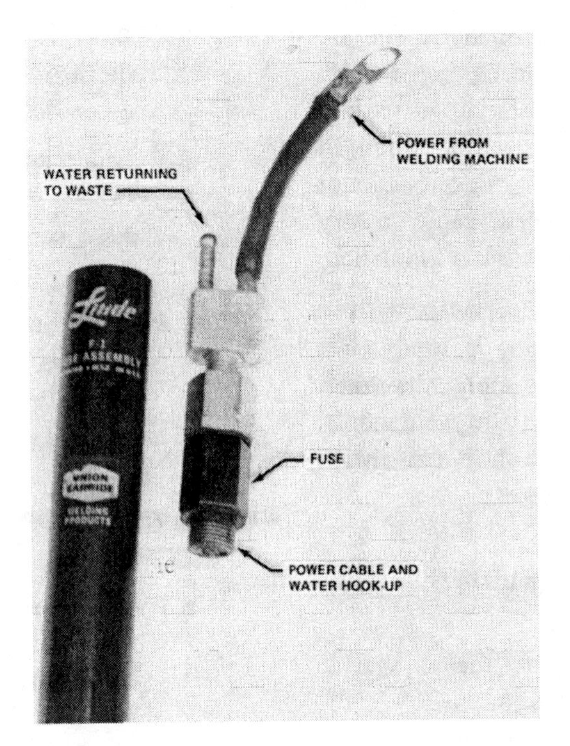

Fig. 9-8 TIG welding torch and cable protection fuse.

7.3 Exhaust plasma cutting fumes, especially when cutting stainless steel.

7.4 The plasma torch leads should be kept up so that they are not damaged by sparks or stepped on.

7.5 Power cut-off devices should be in the water line to prevent the torch from being used without cooling water flowing, figure 9-8. Without cooling water, the torch can burn up and short out.

REVIEW QUESTIONS

Read each question and indicate the number of the rule in the unit which most accurately answers the question.

1. Why is is important to stop water leaks on a G.T.A.W. torch?
2. What can happen if a stud welding gun is not held squarely on the work?
3. What can happen when spatter builds up on a G.M.A.W. nozzle?
4. Why don't you wrap welding leads around your arm to relieve the weight for welding?
5. When should a lifeline be used?

Read each question and indicate the letter next to the statement that most accurately answers the question.

1. S.M.A.W. electrodes should be used until they are approximately how long?

   a. less than 1 inch (25 mm)
   b. 1 to 2 inches (25 mm to 50 mm)
   c. 1 1/2 to 2 inches (38 mm to 50 mm)
   d. 2 to 3 inches (50 mm to 75 mm)

2. Why should you be careful when you put down a G.M.A.W. gun?

   a. the trigger may be accidently pushed
   b. it may short out
   c. the cup will break easily
   d. all of the above

3. What type of gloves should be used for G.T.A.W.?

   a. deerskin gloves
   b. light cotton gloves
   c. gauntlet-type leather gloves
   d. either a or b

4. What poisonous gas may be present during G.M.A.W. welding with carbon-dioxide shielding gas?

   a. ozone
   b. carbon dioxide
   c. carbon monoxide
   d. all of the above

5. Why is it important to have a solid footing and something to hold onto when you have the welding hood down?

   a. the hood can easily fall off
   b. it is easy to lose your balance
   c. someone may pull your cable by accident
   d. all of the above

Unit 10 Equipment

OBJECTIVES

After completing this unit, the student will be able to explain how to correct any safety defects found on arc welding equipment and plan for the safe location of arc welding equipment.

The possibility of the welder receiving an electric shock has been minimized by machine design. However, proper installation, safety checks, and routine maintenance are needed to keep this danger from increasing through equipment failure over time.

1. LOCATING AND INSTALLING EQUIPMENT

1.1 Locate equipment so that water will not stand around the machine base. A wet machine must be thoroughly dried and tested before it is used.

1.2 Locate welding leads well away from main power leads, so they cannot come in accidental contact with high voltage.

1.3 Cooling water lines should have a drip loop so that leaks and condensation will not run back in the machine, figure 10-1.

1.4 Locate main power lines overhead and drop them to each machine location.

1.5 Locate equipment so that it will not be tampered with by unauthorized personnel.

1.6 Locate equipment power disconnects so they can be reached quickly in an emergency, without endangering the person who disconnects the power.

1.7 Locate welding equipment well away from overhead cranes or work aisles. This minimizes the chance of danger to the equipment.

1.8 Welding machines should be located so that they are not exposed to corrosive fumes, welding sparks or excessive dust.

1.9 Ground all equipment frames or cases.

1.10 Provide fuses or circuit breakers for overload protection. They must be properly sized for the machine current requirements.

1.11 Magnetic safety interlocks can be provided to prevent equipment from restarting after a power failure.

1.12 Main power terminals must be located inside the welding machine cover and must be accessible only with tools.

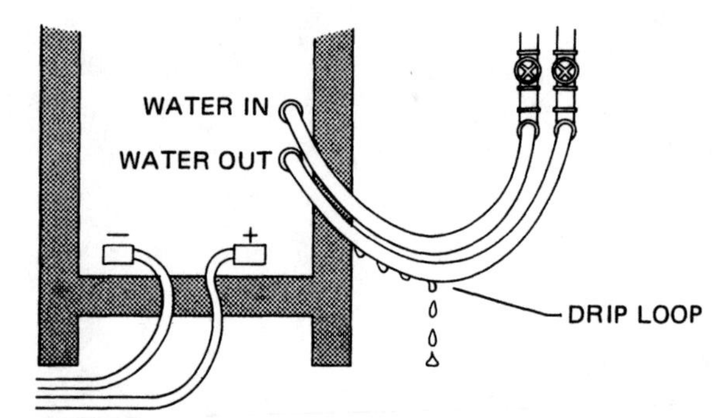

Fig. 10-1 Note that the water hoses have a low spot between the supply and the machine. Condensation and leaks will drip well away from the machine itself.

1.13 Welding cables must not have a splice within 10 feet (3 m) of the electrode end.

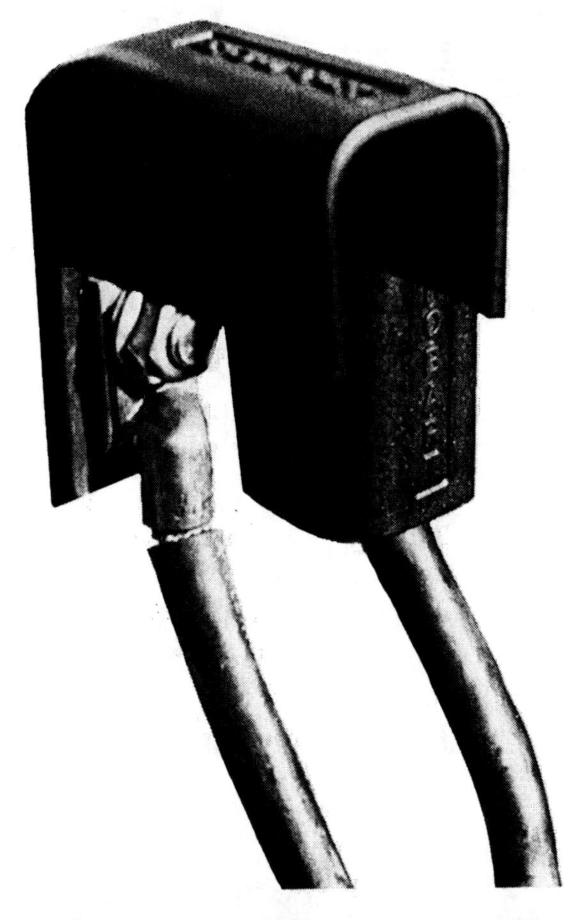

Fig. 10-2 Some power-lug protection is provided by add-on boots (Miller) and some machines are provided with it (Hobart).

1.14 Provide covers for the lug terminals so that they cannot be accidently shorted out by a wrench, welding wire, or any other metal object, figures 10-2a and 10-2b.

1.15 Do not use piping or building structures as a permanent work ground. They can be used as a ground when they are being welded.

1.16 Do not use chains, hoists, or elevator cables as welding grounds.

1.17 Welding leads should run above walkways or be covered so that they are not a hazard to workers figure 10-3.

1.18 An open circuit with no load voltage should be no higher than 80 volts on A.C. machines and 100 volts on D.C. machines.

1.19 Portable plug-in controls operated by the welder must operate on 120 volts or less.

1.20 G.T.A.W., or any other welding leads that carry high frequency for starting the arc, must be 25 feet (7.6 m) or less in length. If they are over 25 feet (7.6 m), they can become a transmitting antenna and interfere with some electrical equipment.

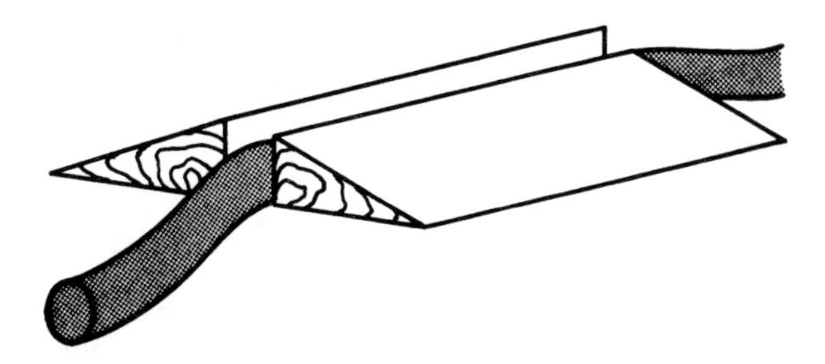

Fig. 10-3 Two blocks of wood can be laid beside cables to prevent people from tripping.

1.21 Welding leads should only be as long as needed for the job.

2. SAFETY CHECKS AND MAINTENANCE

2.1 Check all cables and wires for damaged insulation. Repair or replace as needed.

2.2 Check and repair fuel and coolant leaks on portable engine generators.

2.3 Disconnect power. Remove machine covers and blow dust and dirt from all internal parts, figure 10-4.

2.4 Check welding-lead terminal lugs for tightness. Clean any corrosion.

2.5 Turn off malfunctioning equipment or equipment causing electrical shock. Report this to your supervisor or instructor.

2.6 Do not use damaged or broken equipment.

2.7 Electrode holders that become hot should not be cooled in water, but should be cleaned or replaced.

2.8 Lubricate moving parts with approved greases and oils.

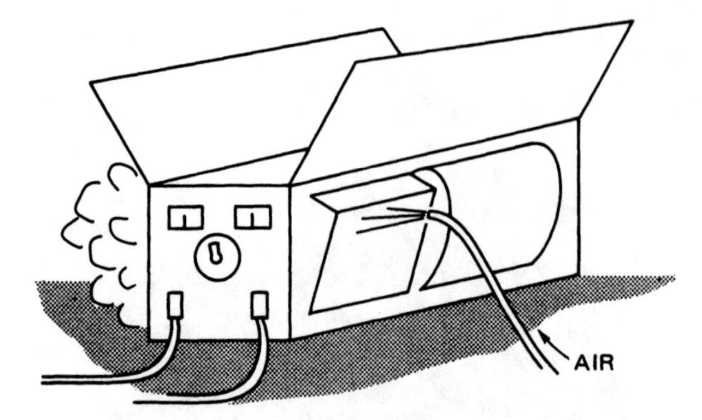

Fig. 10-4 Slag, chips from grinding, and dust must be blown out occasionally so that it will not cause a fire or short out the machine.

2.9 Splice welding cable with approved connectors only and insulate it properly.

2.10 If two or more sizes of cables are to be spliced, the largest cable should be located closest to the machine and the smallest cable at the work end.

2.11 Welding-torch cooling water should have a temperature rise of 10 to 15 degrees Fahrenheit (5 to 8 degrees Celsius), or less. A higher rise may cause the water flow to be restricted and the torch head to overheat.

REVIEW QUESTIONS

Read each question and indicate the number of the rule in the unit which most accurately answers the question.

1. How many feet from the electrode end of a welding cable can a splice be?
2. What should you do if an electrode holder becomes hot?
3. How should equipment power disconnects be located?
4. Why should welding leads be located well away from main power leads?
5. When 2 or more sizes of welding cable are to be used, how should they be arranged?

Read each question and indicate the letter next to the statement that most accrately answers the question.

1. Why are welding-cable terminal lug covers used?

   a. to relieve the strain on cables
   b. to prevent accidental shorting out by a wrench or other tool
   c. to keep the lugs clear of corrosion
   d. none of the above

2. Welding torch cooling water should have how much of a temperature rise?

   a. 10°F to 15°F (5°C to 8°C)
   b. 5°F to 25°F (2.5°C to 14°C)
   c. 18°F to 30°F (10°C to 17°C)
   d. under 5°F (2.5°C)

3. What should you do with equipment that causes an electrical shock?

   a. report it to your supervisor or instructor
   b. turn it off
   c. use insulated gloves
   d. both a & b

4. Plug-in controls operated by the welder must have an operation voltage of:

   a. 80 volts AC or less
   b. 100 volts DC or less
   c. 120 volts or less
   d. all of the above

5. If welding equipment gets wet:

   a. dry it out completely before using it
   b. use well insulated gloves when using it
   c. cut off all power while adjusting it
   d. all of the above

CERTIFICATE OF COMPLETION

This Certifies That

has satisfactorily completed the course of Safety For Welders and is therefore entitled to this

Certificate

Given at ______________________________
(school)

____
(date)

____
(instructor)

____
(student)